J. Kegay
April. 2019

The freedom
of scientific research

Manchester University Press

CONTEMPORARY ISSUES IN BIOETHICS, LAW AND MEDICAL HUMANITIES

Contemporary Issues in Bioethics, Law and Medical Humanities, edited by Rebecca Bennett and Simona Giordano, of the University of Manchester, was established to publish internationally respected book-length works – primarily monographs and edited collections, but also specialist textbooks – in contemporary issues in bioethics, law and the medical humanities and as such welcomes proposals from this range of academic approaches to pertinent issues in this area. The series focuses on the strong foundations and reputation of the University of Manchester's world-leading scholars in bioethics and law, and its internationally respected Centre for Social Ethics and Policy. Works from across the humanities, brought to bear on contemporary, historical and indeed future bioethical questions of the highest social and moral concern and interest, will find a perfect home within this series.

FROM REASON TO PRACTICE IN BIOETHICS
An anthology dedicated to the works of John Harris
Edited by John Coggon, Sarah Chan, Søren Holm and
Thomasine Kushner

Medicine, patients and the law: Sixth edition
Margaret Brazier and Emma Cave

Edited by Simona Giordano in collaboration with John Harris and Lucio Piccirillo

The freedom of scientific research

Bridging the gap between science and society

Manchester University Press

Published by Manchester University Press
Altrincham Street, Manchester M1 7JA

www.manchesteruniversitypress.co.uk

British Library Cataloguing-in-Publication Data
A catalogue record for this book is available from the British Library

ISBN   978 1 5261 2767 9   hardback

First published 2019

Typeset
by Toppan Best-set Premedia Limited
Printed in Great Britain
by TJ International Ltd, Padstow

to Nico

Contents

List of figures

List of tables

Contributors

Sean Aas is Senior Research Scholar at the Kennedy Institute of Ethics and Assistant Professor in the Philosophy Department at Georgetown University. His primary areas of research are bioethics, metaethics, and social and political philosophy, with a significant focus on issues of disability: disability as social construct, disability and political egalitarianism, disability and health. These interests tie to broader projects: on the nature and basis of our rights in our bodies; the grounds of egalitarian justice; and the import of diverse embodiment for healthcare ethics and health policy.

Roberto Baldoli is Associate Lecturer at the University of Exeter, and a member of the Centre for European Governance. He received his PhD in politics from the University of Exeter in 2015. His main research interests revolve around theory and practice of nonviolence. In his PhD thesis, he reinterpreted nonviolence as an impure praxis, a non-systematic revolutionary approach aiming at freedom and plurality. Roberto is also interested in the European Union. He has published on different aspects of European politics, such as the European Citizens' Initiative and the Spitzenkandidaten Procedure.

Andrea Ballabeni is Clinical Program Leader at Chiesi Farmaceutici and Visiting Scientist at Harvard T. H. Chan School of Public Health. Ballabeni worked in molecular and cell biology basic research for several years, doing doctoral research at the European Institute of Oncology and postdoctoral research at the Harvard Medical School. He later shifted focus towards clinical and public health research as well as science policy. He then worked as Adjunct Assistant Professor of Natural and Applied sciences at Bentley University, teaching courses on human biology and doing research in the Center for Integration of Science and Industry. He is currently working as Clinical Program Leader at Chiesi Farmaceutici, and collaborating as a Visiting Scientist with the Harvard T. H. Chan School of Public Health (USA) on science policies to increase the quality and the output of scientific investigations. Ballabeni is also a science advocate and has collaborated with Associazione Luca Coscioni since 2006. He holds a PhD in molecular oncology from the Open University (UK), a master's in public health from the Harvard T. H. Chan School of Public Health and an MSc in biology from the University of Parma (Italy).

Andrea Boggio is Professor of Legal Studies at Bryant University. He is the author of *Compensating Asbestos Victims. Law and the Dark Side of Industrialization* (2013) and the co-editor of *Health and Development: Toward a Matrix Approach* (2009). He has extensively written on science policy issues. He earned a doctoral degree from Stanford University and completed his post-doc training at the University of Geneva. He taught at the Centre for Professional Ethics at Keele University (UK).

Iain Brassington is a senior lecturer in the School of Law at the University of Manchester. A philosopher by training, his work covers many areas of applied ethics, with a particular focus on bioscience, and the interaction between ethics and law. His most recent book, *Bioscience and the Good Life*, was published in 2013; he is currently planning a book on the privacy or otherwise of genetic data. Iain is an editor of, and contributor to, the *Journal of Medical Ethics* blog; in 2016, he was invited to sit on the BMA's Medical Ethics Committee.

Marco Cappato is Coordinator of the World Congress for Freedom of Scientific Research, Treasurer of the Luca Coscioni Association and promoter of campaigns for the legalisation of euthanasia and illicit drugs in Italy, also through several civil disobedience actions. Cappato served as an MEP (1999–2004 and 2006–09). He has been European Parliament Rapporteur on: 'privacy in electronic communication Directive'; 'human rights in the world for 2007'; 'production of opium for medical purposes in Afghanistan'; 'public access to EU documents'. He was nominated for the Politician of the Year award organised by *Wired* magazine in 2003 in San Francisco and was winner of the European of the Year award organised by *European Voice* magazine in Brussels.

Gilberto Corbellini is Full Professor of History of Medicine and Bioethics, and Director of the Museum of the History of Medicine in the Department of Molecular Medicine at Sapienza University of Rome. He has published several books, peer-reviewed and newspaper articles about the history of immunosciences, neuroscience, medical ethics and epistemology, and on science and politics. Recent books include: *Perché gli scienziati non sono pericolosi* (2009); *Scienza quindi democrazia* (2011); *Scienza* (2013); *Tutta colpa del cervello* (with E. Sirgiovanni, 2013); *Bioetica per perplessi* (2016). He is working with Elisabetta Sirgiovanni on a new book on the philosophy of evolutionary psychiatry.

Daniela Cutas is Associate Professor of Practical Philosophy at Umeå University as well as at the University of Gothenburg. Her current work focuses on the ethics of close personal relationships, reproduction and parenting, and her research interests are broadly in bioethics and research ethics. She has published in these research areas in journals such as Bioethics, Cambridge Quarterly of Health Care Ethics, Reproductive BioMedicine Online, Journal of Medical Ethics, Health Care Analysis, Science and Engineering Ethics and Hypatia. She is a co-editor of the volumes *Families – Beyond the Nuclear Ideal* (2012) (with Sarah Chan) and *Parental Responsibility in the Context of Neuroscience and Genetics* (2017) (with Kristien Hens and Dorothee Horstkötter).

Davide Danovi leads the Cell Phenotyping Platform within the Centre for Stem Cells and Regenerative Medicine at King's College London. He holds an MD from University of Milan and a PhD in molecular oncology from the European Institute of Oncology. He completed his postdoctoral training working with Prof. Austin Smith and Dr Steve Pollard at the University of Cambridge and University College London. He has several years' experience using stem cells to develop image analysis-based platforms for phenotyping, disease modelling and drug discovery both in academia and biotechnology companies.

Simona Giordano is reader in bioethics at the University of Manchester, School of Law. She is a philosopher by background and is the author of *Understanding Eating Disorders* (2005), *Eating Disorders and Exercise* (2010) and *Children with Gender Identity Disorders* (2012). She edited the first *Anthology on Scientific Freedom* (2012), with John Coggon and Marco Cappato. She is also the author of over sixty scientific publications, and has been working with the Luca Coscioni Association on the ongoing Project on Scientific Freedom since 2006.

John Harris is Professor Emeritus in Bioethics, University of Manchester and Visiting Professor at King's College London. His many books include *On Cloning* (2004), *Enhancing Evolution* (2007) and *How to Be Good* (2016).

Sir Peter Lachmann is Emeritus Sheila Joan Smith Professor of Immunology in the University of Cambridge and a fellow of Christ's College. His principal research interests are: the immunochemistry, biology and genetics of the complement system; immunopathology, particularly in relation to systemic lupus erythematosus and to multiple sclerosis; and insect sting allergy. He was the Founder President of the Academy of Medical Sciences (1998–2002). He served as its representative on the Inter Academy Medical Panel executive (2000–06). He has also been Biological Secretary of the Royal Society (1993–98) and President of the Royal College of Pathologists (1990–93), and served on UNESCO's international bioethics committee (1993–98). In these capacities he has become involved with the ethical and policy aspects of medical science, particularly in connection with public health, vaccination, stem cells, transmissible spongiform encephalopathies and genetically modified food crops.

David Lawrence is REA Postdoctoral Fellow at Newcastle University Law School (UK). He received his BSc in biomedical science and LLM in bio-technological law and ethics at the University of Sheffield (UK), and his PhD from the Institute for Science Ethics and Innovation of the University of Manchester. His research centres around the ethical and policy implications of emerging technologies, with particular focus on human enhancement, genome editing, and the potential for the development of new types of conscious being.

Heidi Mertes is a bioethicist at the Bioethics Institute Ghent (Belgium), a lecturer in ethics at Ghent University and a postdoctoral research fellow of the Research Foundation – Flanders. She has been affiliated with the department of Philosophy and Moral Sciences of Ghent University since 2005. Research interests include the ethics of embryonic stem cell research, the ethics of fertility preservation, the ethics of prenatal genetic testing and other ethical issues in reproductive medicine. She is a member of the Belgian Federal Commission for Medical and Scientific Research on Embryos *in vitro* and of the European Society for Human Reproduction and Embryology.

Lucio Piccirillo is Professor of Radio Astronomy Technology, School of Physics and Astronomy, University of Manchester. Former Director of the Jodrell Bank Observatory in Cheshire (UK). His main field of research is the development of astronomical instrumentation for astrophysics and cosmology. He is the author of more than 100 publications.

Claudio M. Radaelli is Professor of Political Science, Jean Monnet Chair in Political Economy and Director of the Centre for European Governance at the University of Exeter. A comparative policy analyst, Claudio has published eighty articles and written or edited seventeen books and special issues of academic journals. His main fields of specialisation include the theory of policy learning, Europeanisation, the role of economics in public policy, and regulation. In 2016, Claudio was awarded a European Research Council advanced grant on Procedural Tools for Effective Governance (Protego). In the same year, he edited with Claire Dunlop the *Handbook of Regulatory Impact Assessment*.

Catherine Rhodes is Academic Project Manager, Centre for the Study of Existential Risk at the University of Cambridge. Her work broadly focuses on the interactions between and respective roles of science and governance in addressing major global challenges. She has particular expertise in international governance of biotechnology, including biosecurity and broader risk management issues. She has a background in international relations, but has engaged in extensive interdisciplinary work. Her PhD was funded as part of a Project to Strengthen the Biological Weapons Convention at the Bradford Disarmament Research Centre, and she retains a strong interest in international actions to prevent misuse of bioscience. She worked for the Institute for Science, Ethics and Innovation at Manchester University (2008–15), where her work included: elaborating on the meaning and content of scientific responsibility at the global level; investigation of science advisory processes in international organisations; and a substantial study of the international governance of genetic resources, which has significant implications for the use of biosciences in managing major global challenges.

Cesare P. R. Romano is Professor of Law and W. Joseph Ford Fellow at Loyola Law School, Los Angeles, where he founded and directs the International Human Rights Clinic. He is well known for his work on international courts and tribunals, having published seven books and dozens of papers on this topic. He is co-editor of the *Oxford Handbook of International Adjudication*. He earned a doctoral degree from the Graduate Institute of International Studies, University of Geneva, and has master's degrees from the University of Geneva (in international relations), and NYU (in law).

Gennaro Selvaggi graduated as Medical Doctor, *cum laude*, in 1998 and qualified as Plastic Surgeon, *cum laude*, at the Catholic University in Rome, Italy. As winner of several international prizes, he completed both Reconstructive and Aesthetic Fellowships in London, Dallas, and Gothenburg. He further obtained a PhD in Gender Affirmation Surgery at the University of Ghent (Belgium, 2010); he graduated with distinction as MSc in Leadership and Managing in Healthcare Organisations at the Greenwich School of Management / Plymouth University (2013). Finally, he qualified in 2018 as MA in Bioethics at the New York University. Currently, he is Senior Consultant, and Associate Professor in Plastic Surgery at the Sahlgrenska University Hospital in Gothenburg (Sweden), where he is Project Managing the Program of Gender Affirmation Surgery, and performing this type of surgery. He has presented almost 120 times at international meetings, and published almost 65 peer-reviewed articles in the fields of plastic surgery, gender affirmation, experimental surgery, and clinical ethics. He is on the board of several international journals.

Elisabetta Sirgiovanni is currently Fulbright Research Scholar at the New York University Center for Bioethics. She worked as a research fellow at Sapienza University of Rome and at the National Research Council of Italy. She is Member of the Board of Directors of the Italian Society of Neuroethics.

She has published on philosophy of psychiatry, philosophy of neuroscience and neuroethics. She co-authored *Tutta colpa del cervello* (with G. Corbellini, 2013), which received the National Award for Best Scientific Communication and the Cultural Award 'Mario Tiengo' (2014). She is currently working with Gilberto Corbellini on a new book on the philosophy of evolutionary psychiatry.

Anna Smajdor is Associate Professor of Practical Philosophy at the University of Oslo. Prior to that, she was Ethics Lecturer at Norwich Medical School, University of East Anglia. Anna's research interests incorporate a range of bioethical themes. She has worked extensively on the ethics of new reproductive technologies, and has published widely on medical and research ethics. She is interested in questions concerning the relationship between nature and morality, especially in the context of medicine, scientific research and innovation.

Mary Woolley is the president of Research!America, the US's largest not-for-profit alliance working to make research to improve health a higher national priority. She is an elected member of the National Academy of Medicine (formerly the Institute of Medicine) and served two terms on its Governing Council. She is a recipient of the Adam Yarmolinsky Medal, awarded for exceptional service to the Academy. She is a Fellow of the American Association for the Advancement of Science and served two terms on the National Academy of Sciences Board on Life Sciences. She is a Founding Member of the Board of Associates of the Whitehead Institute for Biomedical Research, a member of the board of the Institute for Systems Biology, a member of the visiting committee of the University of Chicago Medical Center, and a former member of the National Council for Johns Hopkins Nursing. She holds two honorary doctoral degrees, from the Northeast Ohio Medical University (NEOMED) and Wayne State University. She has served as president of the Association of Independent Research Institutes, as editor of the *Journal of the Society of Research Administrators*, as a reviewer for the National Institutes of Health and National Science Foundation, and as a consultant to several research organisations. She has a thirty-year publication history on science advocacy and research related topics, and is a sought-after speaker, often interviewed by science, news and policy journalists.

Series editor's foreword

In 1984 the philosopher and bioethicist John Harris met the theologian Anthony Dyson, by chance on a train journey to a conference they were both attending in Aberdeen. Despite both working at the University of Manchester and both sharing an interest in exploring the moral issues that arise from healthcare and bioscience, they had never heard of each other, so the story goes, until that train journey to Scotland. This chance meeting was the beginning of not only an enduring friendship and academic collaboration between these two men, but also the first step towards the creation of the now famous Centre for Social Ethics and Policy (CSEP) which has been so influential in the area of bioethics, medical law and medical humanities more generally. Together with the lawyer Margaret Brazier and clinician Mary Lobjoit, Harris and Dyson set up this Centre later that year to enable both this new academic alliance and a wider network of academics and students to pursue research and teaching in this area in an innovative interdisciplinary way.

My connection with this ground-breaking interdisciplinary Centre began in 1993 as a junior research fellow on a multidisciplinary, international research collaboration that involved participants from fourteen countries and eleven academic disciplines and so began my apprenticeship in this particular brand of bioethics, law and medical humanities with this amazing group of people and the networks they created. Simona Giordano, the editor of this edition and my co-editor in this book series, joined the Centre shortly afterwards as a research student. It is therefore a particular pleasure of mine to write the foreword for this edited edition that is so very much in the spirit of our Centre and the work that is done there.

This impressive edited volume is the third book in the series I co-edit with Simona Giordano entitled *Contemporary Issues in Bioethics, Law and Medical Humanities*. The first book in this series was an edited volume that celebrated the work of John Harris's, *From Reason to Practice in Bioethics*. The second book was the new edition of the brilliant and ever popular *Medicine, Patients and the Law* by Margaret Brazier and Emma

Cave. As such we are proud that this series reflects the people, the academic rigor and interdisciplinary flavour that has been the essence of our Centre from those very early beginnings. Typical of the CSEP approach, this series not only highlights the work of those directly involved in the Centre but also the research networks that have grown from the work initiated by the Centre and its team including the many students and researchers and other collaborators who have been and remain a part of the CSEP family.

This edited volume *The Freedom of Scientific Research: Bridging the Gap between Science and Society* fits perfectly into this interdisciplinary approach to research into the important area of scientific research and innovation with contributions from an impressive range of disciplines including bioethics, philosophy, life sciences, history, law, social science and medicine. It is this kind of interdisciplinary and international collaboration that has been so central to the success of bioethics and medical law across the decades and is one that we are proud to continue both as a Centre of research and education and as editors of this series with Manchester University Press.

Becki Bennett
Professor of Bioethics
Centre for Social Ethics and Policy
University of Manchester

Preface: the moral foundations of freedom of science[1]

John Harris

This volume of essays under the banner of 'freedom of science' constitutes a new direction for science ethics. One fundamental issue has been and remains defending the idea of the freedom of scientific inquiry and research from political, legal and social restraints. The reasons for maintaining this defence are many. Principal among these are the good that science does, the way it relates to fundamental elements of human nature and the hope it offers to humankind and the planet; more concerning all of these in a moment.

Separating the freedom of scientific inquiry and research from questions concerning the application of science – the progress of discovery, research and innovation through proof of principle to products in the clinic and the marketplace, is fraught with difficulty. For one thing these are often a continuous process and often unstoppable from the perspective of individual jurisdictions. For another, science has increasingly become 'democratised'. In part this has been a deliberate choice with a movement from within science now becoming increasingly conspicuous. This movement calls itself 'citizen science' (Vayena et al. 2016) and involves the encouragement of citizen participation in scientific activity essentially conceived and organised, not ground up by the citizens, but essentially top-down by professional scientists of one sort or another.

A more worrying version of citizen science however has also sprung up and involves the increasing ability of scientifically (often self-) educated citizens creating labs in their garage or kitchen to make a wide range of products free from regulation, or codes of ethics or even conceptions of good or even safe practice (Royal Society 2012; *Scientific American* 2017; Prepperzine 2017; Wikipedia 2017). This sort of citizen science, because it is often secret and always unregulated, gives opportunities for terrorists of all sorts, but particularly for bioterrorists to manufacture weapons formerly required professional expertise and often expensive and conspicuous facilities.

A rather different case, which also crucially engages freedom of science, involves research on human stem cells derived from embryos which is illegal in some jurisdictions and permissible in others (Robertson 2001; EuroStemCell 2008–17; National Conference of State Legislatures 2016). The result is work continues in the UK, for example, which is illegal elsewhere. Scientists move, relatively freely between jurisdictions. What we find is not power without responsibility but responsibility without power. Nations assume responsibility for activities within their borders but turn a blind eye to what their nationals do abroad. The extent to which this is to be celebrated of course varies with attitudes to the substantive ethical issues.

The idea that justice delayed is justice denied continues, rightly, to have currency. But just as justice delayed is justice denied, so therapy delayed is therapy denied, and because illness is confining and health liberating, freedom reappears as inextricably allied to science and medicine. Likewise, 'scientific freedom', freedom to do and publish scientific research, is also often advocated as a basic right (Edsall 1975; Giordano et al. 2012). One reason, to have, not *faith* in science (heaven forbid!), but to put cautious trust in science, is that science has indeed proved to be 'magic that works'. It is the fact that science works, and snake oil does not, that, above all, makes science trustworthy.[2]

Equally fruitless of course is the concentration on protection against real and present dangers, while neglecting preparedness for future threats. Preparedness for the future calls for science and technology and for the habits of mind, free inquiry, reliance on evidence and argument, and above all intellectual honesty, which characterise science broadly conceived.

It is important to remind ourselves of the moral nature of science, threatened, today more than ever, by a culture of reckless deceit, shameless denial of history, and of evidence, and the profligate (Garver 2015; Berrien 2016;[3] Abramson 2017) invention and repetition of more convenient 'alternative facts'. The dishonesty and untruths perpetrated by the culture of alternative facts are polluting every aspect of those freedoms that are worth fighting for.[4] I have been preoccupied with the moral character of science for a very long time (here's why: Harris 1985: especially chs 3, 5 and 6; Harris and Sulston 2004; Harris 2005; Chan and Harris 2009; Chan et al. 2010; Harris 2013).

We all benefit from living in a society, and, indeed, in a world in which science is respected and in which science flourishes. Science and the discovery and innovation it generates, resulting in products in the clinic and the marketplace, no less than the objectivity, rigorous analysis, evidence and respect for truth it promotes, is in the interests of us all (see for example Harris 1997; 2005; Zee et al. 2010).

The other *imperative for science (and for philosophy)*

While there are powerful moral reasons for doing science and philosophy, these activities are not necessarily pursued solely (or even principally) for moral, or even for prudential reasons, powerful as these are. There is a simpler, but perhaps even more powerful, imperative at work (Harris 2018).

We humans are curious birds; we like to understand stuff.[5] We like to know *why*, to know *what* and to know *how* and to know *whether*. We like to know how things work, and what they are for, or what they are good for. We also like to know why things happen and the probability of their occurrence. This includes the question of why we exist at all. We spend a lot of time on such things, and we do so, not because it is good for us, or because either the questioning process, or the answers, conduce to our welfare or well-being or make us happy, or protect our vital rights or interests or confer evolutionary advantage (although they may). We do so because that's the sorts of creatures we are: curious birds who like to ask and answer questions.

True, there are myriad 'rewards' for education, science and curiosity, the reason we pursue these, however, if one were needed, is in our will,[6] our free will – it's what we choose to do and how we choose to live. But if the exercise of our curiosity is not honest and evidence based, then the exercise of our will is thwarted, we simply won't find out the *why, what, how* or *whether* . . . to questions we ask. We may get 'answers' but they won't be informative, they will simply deliver lies, fantasies or 'alternative facts'.

As Thomasine Kushner and James Giordano (2017) have argued recently:

> It is important to recognize that sound ethical analysis begins with and proceeds from facts. Facts of the context, circumstance, agents, implements, and actions involved. These facts should not be 'alternative', they need to be real. But this is an age of increasing misinformation.

We have been talking about the sorts of creatures we are. But 'we' may be on the verge of creating new unprecedented creatures, not only with powers and capacities comparable to ours, but maybe enhanced beyond those that humans have yet attained, or even beyond those which creatures constituted as we are, with our evolutionary history and maybe also constructed as we are – flesh and blood creatures – can attain. 'We' may soon include both machines and hybrids. But the success of such creations will depend vitally on the nature of the creatures we create and how that nature can develop and relate to or coexist with our own.

If we create beings as smart, or smarter, than us, how can we limit their power to act detrimentally towards us, perhaps deliberately to destroy us, or simply to act in ways that will have this result? Martin Rees (2003) has observed that there may be scientific facts that will never be discovered by beings with brains that have evolved in the way that human brains have so far developed, and scientific theories creatures with our evolutionary history are incapable of postulating. One reason for creating AI persons might then be to solve problems we humans cannot address or even imagine.

How can we ensure that such creatures, if we bring them into being, will act for the best? Some have thought that this problem can be solved by programming them (or us) to obey some version of Isaac Asimov's so-called 'laws' of robotics, particularly the first law: 'a robot may not injure a human being, or, through inaction, allow a human being to come to harm'. The problem of course is how the robot would be able to obey such a law when ethical dilemmas often involve choosing between greater or lesser harms or evils rather than avoiding harm altogether; or by allowing or causing some to come to grief for the sake of saving others. How would they be able to keep their eyes on the protective prize?

The question of how to combine the capacity for good, with the freedom to choose is probably one of the things Stephen Hawking had in mind when he told the BBC in 2014 that 'the primitive forms of artificial intelligence we already have, have proved very useful. But I think the development of full artificial intelligence could spell the end of the human race' (Cellan-Jones 2016). How might AI persons, who could determine their own destiny, as we humans do, be persuaded to choose modes of flourishing compatible with those of humans? Of course we currently have these problems with respect to one another; but at least we have not as yet shackled our capacity to cope with these by creating AI persons which may be 'programmed' in ways that selectively preclude acting on the basis of genuine choices informed by evidence and argument (Harris 2011; Harris 2014; Palacios-González and Harris 2014).

As we emerge into a post-truth fantasy world, a Trumped-up world of lies and 'alternative facts' this problem becomes acute. In such a world how can there be genuine choices informed by evidence and argument? This post-truth world raises very real questions about the possibility of our long-term survival, either as the sorts of rational moral beings evolution has painstakingly made us, or indeed as beings of any description at all.

Initial scientific predictions on the survival of our planet suggested we might have 7.6 billion years to go before the earth gives up on us. These were Steven Hawking's calculations, but recently Hawking revised his prognosis: 'I don't think we will survive another thousand years without escaping beyond our fragile planet' (Cellan-Jones 2016). And Martin Rees (2003) has speculated that this might be our 'final century'.

In view of threats like these, we need to make ourselves, humankind, smarter, more resilient and more aware that honesty, truth and objectivity are not optional and dispensable extras. And we may need to call AI persons in aid to achieve this if we are to be able to find another planet on which to live when this one is tired of us, or even perhaps develop the technology eventually to construct another planet. To do so we will have to change, but not, we may hope, in ways that risk our freedom, our capacities to choose both how to live and the sorts of lives we wish to lead; and also by making sure we avoid the creation of machines who might choose to be our masters.[7]

These are some of the ethical challenges that are created by science and our freedom (indeed our fate) to pursue truth, facts and evidence by, inter alia, various sorts of scientific method. This pursuit has become more urgent in view of increasing awareness of the dangers that threaten humanity and indeed our fragile planet and because of the increasingly parochial decisions made recently in many democracies across the world.

A book like this is thus particularly important at this point in history. Its multidisciplinary contents represents its greatest strength; contributors from several disciplines discuss various areas of scientific research, make it accessible to the non-specialised audience but also engage with the broader question of how regulation can promote and hinder a progress of science that can yield significant benefits for ourselves, the future generations and possibly other animals too, indeed providing the human species with a concrete hope for survival. But the other important aspect of this book is *how* it came to exist. Many of the contributors to this volume have been engaged in an ongoing forum for over a decade now, participating over the years in a regular arena of debate, update and discussion, and this book brings some of these discussions, with different spirit, tone and aims, to a broader audience, in this way concretely bridging the gap between science and society.

Notes

1 I outlined this imperative, inter alia, in Harris (2007: ch. 11) on which I draw here. Also, I freely acknowledge and deploy arguments developed in Harris (2018).

2 Other reasons are its openness, its publication of results for further scrutiny, its rigorous peer-review process, and the fact that good science can only be pursued in free societies. I do not of course have room here to justify these claims.

3 There are many more apparent examples of Trump's alternative facts listed at the sites (listed in the references), but I should warn fellow scientists that I have not myself personally checked any of these, either for accuracy or coherence.

4 I am grateful to Tomi Kushner for a stimulating correspondence on the subject of alternative facts.

5 For a recent 'take' on curiosity, see Kahan et al. (2017). See Harford (2017) for a fascinating account both of the mechanisms and history of alternative facts, but also of the importance of human curiosity as an antidote.

6 Julius Caesar, in Shakespeare's play of that name, justifies his decision (which he later reverses) not to attend the Senate on the Ides of March thus: 'The cause is in my will: I will not come' (Act II, Scene ii).

7 In the following paragraphs I draw on work published with my colleagues in Lawrence et al. (2016).

References

Abramson, J. (2017), '"Alternative facts" are just lies, whatever Kellyanne Conway claims', *The Guardian*, www.theguardian.com/commentisfree/2017/jan/23/kellyanne-conway-alternative-facts-lies (last accessed 26 October 2017).

Berrien, H. (2016), 'Lyin' Donald: 101 of Trump's greatest lies', *Daily Wire*, www.dailywire.com/news/4834/trumps-101-lies-hank-berrien#exit-modal (last accessed 26 October 2017).

Cellan-Jones, R. (2016), 'Stephen Hawking – will AI kill or save humankind?', *BBC News*, www.bbc.co.uk/news/technology-37713629 (last accessed 26 October 2017).

Chan, S., and Harris, J. (2009), 'Free riders and pious sons: why science research remains obligatory', *Bioethics*, 29.3 161–71.

Chan, S., Sulston, J., and Harris, J. (2010), 'Science and the social contract: on the purposes, uses and abuses of science', in J. Billotte and M. Cockell (eds), *Rising to the Challenge of Transdisciplinarity: Tools and Methodologies for a World Knowledge Dialogue*, Lausanne: EPFL Press, 45–61.

Edsall, J. T. (1975), *Scientific Freedom and Responsibility: A Report of the AAAS Committee on Scientific Freedom and Responsibility*, Washington, DC: American Association for the Advancement of Science, www.aaas.org/sites/default/files/SRHRL/PDF/1975-ScientificFreedomResponsibility.pdf (last accessed 7 March 2017).

EuroStemCell (2008–17), 'Regulation of stem cell research in Germany', www.eurostemcell.org/regulation-stem-cell-research-germany (last accessed 26 October 2017).

Garver, R. (2015), 'Donald Trump's 8 (most recent) blatant lies', *Fiscal Times*, www.thefsiscaltimes.com/2015/11/24/Donald-Trump-s-8-Most-Recent-Blatant-Lies. and https://video.search.yahoo.com/yhs/search?fr=yhs-adk-adk_sbnt&hsimp=yhs-adk_sbnt&hspart=adk&p=Donald+Trump+Lies#id=8&vid=2abf45f7778c07d (last accessed 1 March 2017).

Giordano, S., Coggon, J., and Cappato, M. (eds) (2012), *Scientific Freedom: An Anthology on Freedom of Scientific Research*, London: Bloomsbury.

Harford, T. (2017), 'The problem with facts', *Financial Times*, www.ft.com/content/eef2e2f8-0383-11e7-ace0-1ce02ef0def9 (last accessed 24 April 2017).

Harris, J. (1985), *The Value of Life*, London: Routledge.

Harris, J. (1997), 'The ethics of clinical research with cognitively impaired subjects', *Italian Journal of Neurological Sciences*, 18.5: 9–15.

Harris, J. (2005), 'Scientific research is a moral duty', *Journal of Medical Ethics*, 31.4: 242–8.

Harris, J. (2007), *Enhancing Evolution*, Princeton, NJ: Princeton University Press.

Harris, J. (2011), 'Moral enhancement and freedom', *Bioethics*, 25.2: 102–11.

Harris, J. (2013), 'In search of blue skies: science, ethics and advances in technology', *Medical Law Review*, 21.1: 131–45.

Harris, J. (2014), 'Taking liberties with free fall', *Journal of Medical Ethics*, 40.6: 371–4.

Harris, J. (2018), 'The chimes of freedom: Bob Dylan, epigrammatic validity and alternative facts', *Cambridge Quarterly of Healthcare Ethics*, 27.1, forthcoming.

Harris, J., and Sulston, J. (2004), 'Genetic equity', *Nature Reviews Genetics* 5.10: 796–800.

Kahan, D. M., Landrum, A., and Carpenter, K. et al. (2017), 'Science curiosity and political information processing', *Political Psychology*, 38.S1: 179–99.

Kushner, T., and Giordano, J. (2017), 'Neuroethics: cashing the reality check', *Cambridge Quarterly of Healthcare Ethics*, 26.4: forthcoming.

Lawrence, D. R., Palacios-González, C., and Harris, J. (2016), 'Artificial intelligence – the Shylock syndrome', *Cambridge Quarterly of Healthcare Ethics* 25.2: 250–61.

National Conference of State Legislatures (2016), 'Embryonic and fetal research laws', www.ncsl.org/research/health/embryonic-and-fetal-research-laws.aspx (last accessed 26 October 2017).

Palacios-González, C., and Harris, J. (2014), 'How narrow the strait: the God machine and the spirit of liberty', *Cambridge Quarterly of Healthcare Ethics*, 23.3: 247–60.

Prepperzine (2017), '10 kickass weapons you can make at home', http://prepperzine.com/10-awesome-weapons-can-make-home (last accessed 26 October 2017).

Rees, M. (2003), *Our Final Century*, London: Arrow.

Robertson, J. A. (2001), 'Human embryonic stem cell research: ethical and legal issues', *Nature Reviews Genetics*, 2.1: 74–8, www.nature.com/nrg/journal/v2/n1/full/nrg0101_074a.html (last accessed 26 October 2017).

Royal Society (2012), 'H5N1 research: biosafety, biosecurity and bioethics', https://royalsociety.org/science-events-and-lectures/2012/viruses (last accessed 26 October 2017).

Scientific American (2017), 'Citizen science', www.scientificamerican.com/citizen-science (last accessed 26 October 2017).

Vayena, E., Brownsword, R., and Edwards, S. J. et al. (2016) 'Research led by participants: a new social contract for a new kind of research', *Journal of Medical Ethics*, 42.4: 216–19.

Wikipedia (2017), 'Bioterrorism', https://en.wikipedia.org/wiki/Bioterrorism (last accessed 26 October 2017).

Zee, Y. K., Chan, S., and Harris, J. et al. (2010), 'The ethical and scientific case for phase 2C clinical trials', *Lancet Oncology*, 11.5: 410–11.

Acknowledgements

I wish, first of all, to thank John Harris and Lucio Piccirillo. Without their support, this volume would not have come to exist. My gratitude also goes to Marco Cappato and the whole Luca Coscioni Association, for having included me in all the activities of the World Congress on Freedom of Scientific Research and in many other initiatives, but also for rendering me more alert to the importance of freedom for many of us, and thus more sensitive to the urgency of defending greater freedom of science and of access to medical treatments and scientific innovation. A special thank you to all contributors, who have been incredibly punctual and precise in their job and have made editing this volume very pleasant.

Many thanks to SØren Holm, who has read and commented on some parts of this work.

Part I

Freedom of science: promises and hazards

Introduction to Part I

Simona Giordano

The weight, value and transformative effect of scientific research are greater now than they have ever been. The nature of the moral scepticism that underpinned much late twentieth-century liberalism now bows to a scientific culture, where scientific method and the reliability of peer-reviewed results bear directly on our ethical norms, our national and supranational laws, and on social activities as diverse as farming, medicine, insurance and city planning. Science has thus changed the world, and has changed, as some of the authors discuss in this collection, even our cognitive and moral abilities.

The idea of writing this book was formed a long time ago, in April 2014, after the Third World Congress on Freedom of Scientific Research, held in Rome and organised and sponsored by the Luca Coscioni Association. The editor of this collection, John Harris and Lucio Piccirillo have collaborated with the Luca Coscioni Association and participated in the conference, either as speakers or organisers. But the origins of this book are even older.

The World Congress on Freedom of Scientific Research is an international ongoing forum, which was formed in 2006 in response to concerns in the international scientific community that scientific freedom might be hindered by ideologies that do not stand up to moral or rational scrutiny. In the early 2000s, part of the international scientific and bioethics community was responding with profound concern to innovations in embryological science; the European Union decided to take time to think about the matter, and first imposed a moratorium, and then a series of limits to the funding of scientific research involving human embryos. A heated debate followed regarding the likely repercussion upon the development of regenerative medicine.

Marco Cappato, in Part II of this volume (Chapter 12), discusses the international reaction to 'human cloning' (more properly, cell nuclear transfer) following the announcement of the birth of Dolly the sheep in 1997. In response to the virtually unanimous ban on 'human cloning', and on the restrictions imposed all over the world upon stem cell research and, consequently, upon regenerative medicine, the United States Coalition for the Advancement of Medical Research, the Genetics Policy Institute, together with several Nobel Laureates and a number of representatives of patients' associations, scientists and politicians, including Members of the European Parliament, united in The World Congress for Freedom of Scientific Research.

This book makes the results of this intellectual enterprise available to the international academic community and to the general reader, bridging the gap between communities that often work in isolation from one another: the scientific community, the political community and the academic community. But we aim at making the discourse on the politics of science available and accessible to the general public, attempting to explain how decisions that affect us all are taken, and thus how science and politics function in contemporary society, and perhaps how they could and should function.

This book includes contributions by some of those who participated in the last Congress, as well as contributions by others who joined later. Obviously, the original papers have substantially changed, because a lot has happened since 2014. The focus of this collection is on the relationship between science and society, and its mediation through law, ethics and social, political and economic norms. The authors have the most diverse backgrounds, and therefore their style is diverse, and this comes across clearly; some of the contributors are scientists, others are philosophers, others are politicians or humanitarian activists; their nationalities are also different – some are European, some are not. So the way they convey their message and their writing style differ significantly, and it is hoped that this diversity will contribute to make this collection valuable and original. This book takes a multidisciplinary approach to the problems of scientific freedom. The methodologies used will therefore be those of discursive research in sciences, politics, law, philosophy and economics.

The notion of freedom is central to this book. This notion has evolved historically and has been debated widely over the history of Western thought (Arendt 1993). Depending on the context we may talk of 'freedom of the will', 'legal freedom', 'economic freedom', 'religious freedom' and so on. It may be possible, however, to distinguish or identify four broad meanings of the term 'freedom': *metaphysical freedom*, *negative freedom*, *positive freedom* and *civil and political freedom* (or *freedoms*). This distinction is approximate, but it may be useful to identify the uses and meanings that tend to recur most often in the volume.

In the first sense (which I called metaphysical freedom), freedom refers to the ability of humans to act according to their own will (Mele 1995). This sense of freedom appears relatively recently in the history of Western thought. In Greek mythology and in a large part of ancient Greek philosophy humans are not 'free' in the way we would consider ourselves free today. In ordinary language, to say that I am free may mean, for example, that I am not enslaved, or very simply that I have the ability to choose, say, to eat pasta or chilli con carne, or to go for a walk or watch a film. For many Greek philosophers, this freedom is an illusion. It is 'fate' that determines our choices and actions. Even the gods are subject to fate. Gods and humans may have the sensation of being free to choose, but their life, death and destiny are predetermined. Thomas Hobbes observed that the notion of freedom for most Greek philosophers was a feature of the state (the polis), and was not a feature of individuals (Hobbes 1651: X, 8). For the Stoics, for Heraclitus and Parmenides, men were free insofar as they were able to accept their own destiny (Palmer 2013). A person is like a dog tied to his chain: he can freely run around, and enjoy his freedom, but only insofar as he stays within the length of his chain.

The first known discussions of freedom as we may intend it today, as free will and self-government, are found in the Sophists, and a precursor of this understanding of freedom can also be found in Plotinus, one of the most prominent scholars of Plato, who became a famous philosopher in Rome in his later years. The ideas of free will and moral responsibilities that he contributed to shape were later elaborated and became central to Christian theology. Most of the Western philosophers we know of – Leibniz, Voltaire, Spinoza, Hume, Locke, Condillac, Kant, Hegel, Kierkegaard, Marx, Jaspers, Sartre, to mention just a few – have developed theories of human freedom, intended as our ability to act according to our will, or our intellect, or our reason.

Although the other three meanings of 'freedom' will be more relevant to the contributions in this book, it should be noted that science is relevant to the understanding of freedom in the metaphysical sense. Neuroethics, for example, attempts to evaluate whether individuals form genuinely autonomous decisions, and thus also whether they should be held responsible for their choices and actions. The relevance of this to a number of contexts is remarkable: think of criminal liability. If people are not free in the relevant sense, how can they be held accountable and punished for their actions? Think of informed consent. Are there factors that render our choices non-autonomous? Some religious and cultural influences are regarded as rendering people non-autonomous, and thus as invalidating consent; this is why in the UK for example the Female Genital Mutilation Act 2003 prohibits excisions of labia majora, minora and clitoris for non-medical reasons, even if requested by adult consenting women. Yet in other cases religious beliefs are not regarded as invalidating people's autonomy and their ability to give or withdraw consent to medical procedures – and this is why Jehovah's Witnesses are allowed, in England and in many other countries, to refuse whole blood products, even if they will die as a result of their refusal.

There is another way in which this metaphysical sense of freedom features in this collection. As Corbellini and Sirgiovanni discuss (Chapter 13), science has an impact on our freedom, intended in this first metaphysical sense. The authors argue that the wider accessibility of scientific education has enhanced people's ability to evaluate facts, to reason around them, to make hypotheses, to think abstractly, to think critically and rigorously. These abilities can 'free' individuals from the grip of superstition or from the malice of deceit, as they allow us to reflect critically on what is in front of us, and to question what is said or offered to us. These abilities thus enhance our freedom and moral responsibility, even if we accept that our choices, actions and even preferences are largely determined by factors beyond our control (environment, genes, culture and so on).

Science enables us to know many things about ourselves. This raises other questions around freedom in this metaphysical sense, which are both theoretical and practical, and which are relevant to a number of domains, such as medical ethics, science ethics and even law. Once the information about us, that is precisely about you and about me, is available to you and to me, can we still be free if we refuse to receive it? For example, the sequencing of our whole genome is now possible, and through genetic tests it is possible to evaluate our susceptibility to develop certain diseases, or in rarer cases to establish the presence of genetic mutations that will lead to the development of certain diseases. Can we still be free, in this first sense, if, having this information made available for us, we decided to remain in ignorance? More broadly, what does it take to be free, in this first metaphysical sense? How much knowledge to we need to have about ourselves, how much does 'science' need to disclose about us, for us to be free?

These questions have engaged and divided bioethicists for over a decade now. Harris and Keywood argued that we cannot be free and we cannot make autonomous decisions unless we accept available knowledge about ourselves. Not only can we not be free unless we possess the available information: because without this knowledge we cannot be free, we cannot even freely refuse to obtain that information (Harris and Keywood 2001). Takala and Bortolotti objected that Harris and Keywood misinterpreted here the notion of freedom and autonomy. We can decide not to be told whether we carry a genetic mutation that may or will cause us to develop an illness later in life, and we can still live our life freely (Takala 2001; Bortolotti 2013).

When we apply these considerations to our relationships with children, dilemmas become perhaps even more acute. There has been a debate in the UK as to whether parents who have genetic disorders such as Huntington's chorea (a non-treatable and non-preventable disorder that appears in adulthood, and that is caused by a genetic mutation) could test their children for the disease. The UK Genetic Alliance's recommendation is that children should not be tested for these disorders, because this knowledge violates children's right to an open future. This notion of an 'open future' appears conceptually akin to the notion of freedom in the first metaphysical sense. The UK Genetic Alliance's position is that, as the conditions in questions are not treatable or preventable, knowing about them only limits children's right to an open future (and thus their freedom) (Genetic Alliance 2016).

But, conversely, it could be argued that without that knowledge, freedom is actually taken away from the child. If I have this information about my life as a child, this might allow me to shape my plans and my priorities; I may decide to give precedence to what I can realistically achieve in the time I have at my disposal and avoid long-term plans, for example. Likewise, parents may find that without knowledge there is no open future for their children, only a bet at best, and an inauthentic life based on false hopes at worst. Ignorance may shadow, rather than promote, an 'open future', or our freedom.

The second distinction mentioned earlier, between positive and negative freedom, is usually associated with Isaiah Berlin (Berlin 1982), although John Locke and Jean-Jacques Rousseau had already discussed these concepts. Berlin identified two ways in which the notion of freedom can be understood. One is 'negative': here freedom is freedom *from* (from interference or limitations). If I say, for example, that I am free to marry a person of my choosing, and that nobody should interfere with my choice, I refer to freedom in its negative sense. I am free, and should be free from unnecessary or unjustified external limitations. This sense of freedom echoes John Stuart Mill's notion of liberty: individuals are sovereign over their bodies and their life, and the only purpose for which power can be rightfully exercised over a member of a civilised society against his will is to prevent harm to others (Mill 1859). Modern liberalism is largely based on this view of freedom; individuals can enjoy large spheres of self-government in their private life, insofar as they do not limit the equal freedom of others or do not hurt others.

Positive freedom, instead, is freedom *to*: this freedom usually requires that others provide to me with something so that I can exercise my freedom. If I have a freedom to, say, education, this entails that someone else has a duty to provide me with something so that I can exercise my right (e.g. that the state provides schools). This second sense of freedom echoes discourse of rights: right to education, for example, to work, to life and health.

Positive and negative freedoms are interrelated in many ways. For me to enjoy freedom in the negative sense, it is usually not sufficient that others do not interfere with my choices and actions. Usually it is also necessary that others provide certain things to me. To give an example, suppose that I choose to terminate a pregnancy, and that I live in a country in which this is legally permitted. Negative freedom means that others cannot ethically seize me and force me to keep the baby; they can try to persuade me, to convince me, but they cannot physically interfere with me to prevent me from acting in the way I choose. But in order for me to exercise my freedom, it is not sufficient that others do not interfere with me directly: it is also necessary that the healthcare system provides accessible services. If, say, in rural areas abortion clinics are not available, and women do not have the resources to access available centres, then women are not free, even if their negative freedom is respected.

This is important to the purposes of this collection. If individuals have a right to life and to health, as protected and defended in virtually all declarations and conventions of human rights, this means of course that they ought not to be deprived of their life (unless perhaps they so wish – as in cases of assisted suicide or euthanasia); and of course it means that they should not be exposed to preventable harm, diseases and illnesses (they should not be deliberately infected with transmissible diseases, for example, and they should not be physically assaulted or harmed). This is why many liberal states recognise forms of liability for murder and physical assault, but also for accidents recklessly caused to others. This is also why many liberal states adopt routine or compulsory vaccination programmes, and why in many liberal states healthcare services are made available to all citizens. The provision of these services may cause some restrictions of other freedoms we may also enjoy: for example, the provision of publicly funded health services causes people to pay taxes, and thus limits their freedom to dispose of all their earnings. Some restrictions are usually regarded as proportionate and justified, because of the good that they protect and promote. Your right to life is more important than my claim to be able to drink and drive. It could be argued that if this is true, then limiting the ability of scientists to pursue research into certain areas of medical science is limiting the right to life and to physical integrity of those who would benefit from this research. This is an argument that the reader will find in this volume (see, for example, Chapters 11–12). It makes no sense to speak of negative freedom unless certain barriers that can limit the enjoyment of that freedom are removed. Thus, one could argue that if, for example, stem cell research offers the prospect of treatment for spinal cord injuries, there is little point in saying to a person who is paraplegic that she has a right to non-interference, if the parliament of her country prohibits stem cell research, or if funding for that research is not made available. To say this, it could be argued, is similar to saying to her that we respect her negative freedom to non-interference, but then we do not provide her with a wheelchair. It could be argued that if it is cruel to say to this person that she can rightfully exercise her freedom from interferences while at the same time denying her the available wheelchair, it is similarly cruel to prevent scientific research that is likely to lead to the discovery of treatment for her condition.

Thus, it becomes clear that what it takes to respect people's negative freedom relates very closely to what should be provided for them so that they can actually exercise their freedom. And it is here that complexities arise in the context of freedom of scientific research; in this context the values, priorities and demands are multiple and diverse; how these should be balanced and ranked, and how our most fundamental freedoms should be protected, in order for them not to be mere unfulfilled words, is central to this collection.

In the fourth sense, freedom can be intended as civil and political (in this sense it is more appropriate to talk about *freedoms*). In this sense we refer to freedom from oppression, freedom from coercion, freedom of association, of speech, of movement, of the press; civil and political freedoms are those that many of us relate to as what citizens of liberal democracies enjoy, as opposed to the limitations that characterise totalitarian regimes.

Civil and political freedoms are those that a state cannot legitimately restrict without good reason. Most contemporary states have documents (in the form of a constitution or bill of rights) which state what the basic civil and political freedoms of citizens are. There are also supranational documents, which have similar content, and which can be ratified by individual states, such as the European Convention on Human Rights, or the International Covenant on Civil and Political Rights; these will be discussed in various chapters in this collection.

Although this book is not primarily a book of political philosophy, it is still a book about freedom; in particular, the notions of negative and positive freedom and civil and political freedoms will recur in the works presented here. The notion of metaphysical freedom will probably be less prominent. Freedom of scientific research can be understood as negative freedom (questions here concern the degree of freedom from interference that scientists should enjoy) and as positive freedom (questions here concern the infrastructures and legislative frameworks that should be provided in order for science to operate). Scientific freedom can also be understood as a particular type of civil and political freedom, or as an enterprise that has direct impact upon people's civil and political freedoms.

The support and limits that should be given to science require constant evaluation: priorities need to be set, scarce resources need to be allocated, competing principles and faiths need to be accommodated, obligations and responsibilities need to be distributed among societies' members. Questions about freedom of scientific research are also questions about how free science truly is or can be (Vattimo and Cavalli Sforza 2006). Science is an enterprise, and as such it is directed (at least to an extent) by the political agenda, which, in turn, also determines how funding is allocated. Even in liberal democracies, where parliaments and governments are democratically elected, political agendas do not always reflect the priorities of the people; but of course it may be debatable whether or not it is the priorities of the people, even of the majority of the population, that should steer political agendas and scientific research.

The ethical properties of science are inherently subject to controversy and debate. As we will see in this volume, some areas of biomedical science are by some considered outright wrong. But on the other hand, other areas of research that may appear morally neutral can still be used for morally dubious purposes (see Chapter 6 on bioterrorism in this volume). Other areas of science and technology may be seen as morally neutral, but may be expensive and may not promise immediate 'returns'. Questions may thus be asked about whether it is ethical to invest in these areas of science, in a context of limited resources.

This book will not offer a coherent or conclusive notion of freedom. It rather wants to promote a space for cultural exchange – on paper of course, just with a book – and critical reflection on issues that concern scientific research, its boundaries and who should be setting those boundaries. This collection tries not to be skewed in one direction, but we recognise that a liberal, progressive spirit has moved the World Congress and thus also inspires this book. However, we don't wish to indoctrinate the reader – we believe that within the spirit or culture that has inspired this collection it is possible to reason about the advantages and disadvantages of certain scientific developments and about certain regulatory mechanisms. We recognise that we are probably all somehow 'indoctrinated', whether or not we are willing to admit it. But our aim is to promote a debate, which of course moves from a certain perspective, and which sometimes proposes a certain point of view, but which wants to remain informed and responsible. With this in mind, we have decided not to attempt to level either the style or the voice of different contributors, because we wanted this collection to reflect the pluralism that inspires the enterprise of the World Congress of Scientific Freedom. Our aim is to provide and show the value of different intellectual and practical endeavours, and not to yield a unified message to the reader.

Part I of this collection discusses some of the ways in which science is changing the world. The first two chapters discuss the impact of science (particularly immunology) on human life. Sir Peter Lachmann (in Chapter 1) provides a fascinating overview of the milestones in the immunological sciences and the effect these have had on the duration and the quality of human life. Overall, humankind has lived longer and better since at least the 1900s. But 'all that glitters is not gold', wrote Shakespeare in *The Merchant of Venice*. And Lachmann concludes by unveiling the other side of the coin: the growth of world population is simply unsustainable. CO_2 emissions and indiscriminate use of scarce resources are likely to put humankind at new risks of global deaths and even extinction. Life extension and overall population growth call humans to new levels of responsibility towards the environment and towards each other. These new and more demanding levels of responsibility are the price we need to pay for our longer and healthier lives, and they are the only way to ensure that what is a blessing to many does not become a curse for the generations to come.

My contribution (Chapter 2) continues the discussion on the curses and blessings of scientific progress. I take a different angle, though, and consider the challenges that life extension presents for humankind, particularly in middle- and high-income countries. I consider philosophical and metaphysical concerns around life extension: some thinkers see death, the inevitable death of humans by ageing (not just their inevitable vulnerability to accidents and illnesses), as an inherent feature of humans. The ever-growing ability to delay death, to slow or even reverse the process of ageing, and the possibility of replacing 'old' body parts with younger ones, and even animal or artificial ones, for some changes the very essence of what it is to be human. Some find this trajectory repugnant, unnatural. I propose that life extension is *not* changing the nature of humans – and this is so for the simple reason that it is unclear what this inherent universal and eternal essence of humans is supposed to be; on the contrary, the ability to extend life, or postpone death, are to be celebrated as one of the greatest triumphs of humankind. In practical terms, this means that the growing presence of older people in our societies is also to be honoured.

Similarly to Lachmann, I highlight that social changes are, however, necessary to make population ageing sustainable. These changes include the modification of working patterns, retirement age, city planning and much more. There are also changes in individual lifestyles that need to be responsibly implemented in order to make the most of a long life, and to prevent what is a triumph from becoming the worst of all nightmares for ourselves and those who will come after us. The danger, here, is becoming prey to the old argument that wanted to make people 'responsible' for their own ill health, and thus placed people higher or lower on the scale of healthcare rationing depending on 'how well' they had led their life. The focus, and the responsibility for sustainable population growth, should not exclusively fall upon individuals. There are shared responsibilities, which are social and political, which need to tie in with personal responsibilities.

Science thus changes population structures, demographics, the planet and human life as a whole. But it also changes intimate and private aspects of our lives. Daniela Cutas and Anna Smajdor (Chapter 3) explore some of these transformations in what for centuries have been regarded as natural and thus immutable relationships, namely the relationships between parents and children. In particular, developments in reproductive technologies have created new types of connections between parents and children. Social or legal parenting and genetic parenting have never necessarily coincided, as exemplified by cases of adoption or, traditionally, by cases in which children are raised by, say, grandparents or other members of the extended family. But reproductive technologies, such as *in vitro* fertilisation, allow a woman to give birth to a child who is not genetically *her own*, or to a child who bears genetic material from both herself and another woman (via, for example, mitochondrial DNA transfer). A man can become at the same time a mother and a father (as in the case of the transman who has oocytes harvested prior to transition, or as in the case of the transman who does not seek genital confirmation surgery and thus retains the uterus and ovaries – biologically these people are mothers to their children, but socially and legally they are fathers; or as in the case of people who transition to the other gender after having had children – in these cases they will be biological fathers and social mothers, or vice versa). Women are now able to bear children way past the age of fertility. Perhaps, in the future, a woman will be able to give birth to a child who is genetically *only her own* (via solo reproduction with *in vitro*-created gametes or human cloning). Gamete donation and surrogacy bring about legislative challenges that many legislatures are still trying to sort out. Some people will find these new possibilities marvellous and amazing; some will find them disturbing. Cutas and Smajdor point out that the changes of the last decades are harder and harder to reconcile with the still pervasive *nuclear family* expectations in ethics and regulation. They show how many legislatures across the world still regulate medical care, by limiting access to it, on the basis of an ideology according to which a nuclear family, constituted by a man and a woman, possibly married, conceive children who are genetically related to them. The diversification of genetic parentage itself (alongside a host of other sociocultural changes) pushes the model further into what some may see as the *crisis* of the family.

There is a question to be asked here, namely whether this 'genetic jealousy' has anything to do with the atavistic tendency to tenaciously preserve economic assets within family lineages. It is a question that Cutas and Smajdor do not openly address, and one for which it would be difficult to provide a substantiated answer, but it is a doubt that arises from reading their informative chapter. There is another interesting aspect of the regulation that Cutas and Smajdor highlight: norms and assumptions concerning the structure of the family *constrain* the direction of scientific progress in the area of human reproduction. It is one's family status that determines whether one's reproductive aspirations are classified as medical needs and thus eligible for treatment, or, put differently, as *just personal preferences or desires*. In turn, *needs* (deemed eligible for medical treatment) form the basis of future research priorities. *Desires* do not give impetus to research priorities in the same way. The chapter challenges our assumptions relating to the boundaries of human reproduction, and calls for adjustments in ethics and law to make space for more realistic perimeters of human parenting, and thus for greater freedom of scientific research in the area.

In Chapter 4, Selvaggi and Aas consider another development of scientific research and technology, which raises again a number of issues relating to reproductive ethics, as well as to the ethics of scientific freedom. They focus particularly on recent developments in uterus and penis transplantations. They point out that these types of transplantations, unlike other types, are not primarily meant to save or lengthen the patient's life, but to improve their quality of life by increasing reproductive and sexual function. Also for this reason, they raise questions relating to the ethics of surgical research, when innovative treatment may enhance patients' quality of life if successful, but also expose patients to high risks, including risks of immunological rejection and even death. They go further though; so far, both penis and uterus transplantations have only been performed on cisgender people, that is, people whose gender is congruent with their birth sex. For example, penis transplantations have been attempted on men who have lost their penises due to illness or accidents; uterus transplantations have been performed on women with health problems, who could not bear children. These techniques, however, are in principle viable options for transgender men, that is, women who transition to a male gender, and could represent an alternative to penis reconstruction, which is often unsatisfactory to patients. Uterus transplantations could be utilised to allow transgender women, that is, men who transition to female, to bear children. If these forms of surgery had to be utilised in this population of patients, further challenges, in addition to those highlighted by Cutas and Smajdor, would arise for the law and social policy around who has ethically legitimate claims to reproduce. But Selvaggi and Aas also point out other issues: how are benefits and risks to be judged? And who should make the judgement? Also, importantly, whose organs should be used? In the case of penis transplantations, cadaver organs are normally used – would it be possible to have live donations from transgender women? They discuss the surgical and ethical issues that would need to be addressed in order to answer this question. Finally, they pose a broader question. These are expensive and non-life-saving procedures. Is this a good use of scientific effort?

Another relatively novel development in reproduction is mitochondrial DNA replacement. The issue was in the spotlight in 2015 and 2016 around the world. Iain Brassington explores the ethical issues around this technique in Chapter 5. He explains the technique and its therapeutic goals. Although these are now widely known, there is an interesting aspect of mitochondrial DNA replacement that Brassington examines: namely the relationship between this technique and the broader issue about the ethics of freedom of scientific research. Mitochondrial DNA replacement is in effect a germline modification: an alteration that will be passed down the generations. Hence its promise is potentially double-edged: it may improve future lives significantly over the run of several generations; but if it turns out to have undesirable sequelae, it might cause several generations of harm. This raises questions about the freedom to pursue potentially harmful techniques – not just in relation to this particular case, but across the board. Moreover, some worry that the human genome, or nature, has a value that will be undermined by interferences such as this. But scientific freedom is also valuable. Can we measure one against the other? Finally, what about the freedom *not* to investigate? In a world where the scientific community chose not to pursue such innovations, would that be a proper use of scientific freedom? Brassington argues that scientific freedom includes the freedom *not to pursue research*; in this, he represents a somewhat dissenting voice in the volume. Most of the contributors stress the importance and value of scientific research, in various areas and aspects (Piccirillo considers physics, Mertes and Woolley make quite a strong appeal to ethics in justification of freedom of science, particularly in the field of regenerative medicine; see Chapters 8, 11 and 15, respectively).

Although Brassington notes that there are no particular reasons to be concerned about certain types of germline modifications, such as those brought about by mitochondrial replacements, there is no moral obligation to pursue research, even if it is aimed at preventing potentially serious diseases, such as mitochondrial diseases. Brassington here seems to discount what is known in applied philosophy as the 'equivalence thesis'; according to this thesis, acts and omission may have the same moral weight. There are caveats to the 'equivalence thesis' but at its minimum it states, somewhat convincingly, that harming someone with a positive action is *not necessarily* worse or morally more repugnant than procuring the same harm through an omission. So Brassington leaves unanswered the question of whether deliberately failing to prevent a child from being born in a harmed condition is equivalent to causing that harm. The discourse is complicated further by what philosophers call, after Derek Parfit, the problem of 'non-identity'; in the choice between bringing into the world *two different children*, each child has only got that specific chance of existence. Therefore, if a child is brought into the world with a mitochondrial disease, who would have not been born otherwise, that child is neither harmed nor wronged by being brought into the world. The 'non-identity' problem suggests that a child is not harmed by being brought into existence, even if he or she has a disability or a medical condition, if *he or she could only be born in that state* (unless his or her life is so overwhelmed by suffering that it would be preferable for him or her to have never existed in the first instance).

But there are two problems here. One is this: even admitting that the 'non-identity' problem applies here, one could still say that philosophers seem to be the only category of people who are not concerned about whether a child is born with a serious medical condition or not (provided he or she is not overwhelmed by suffering). The second is perhaps a more compelling problem. As Brassington notes, the mitochondrial DNA represents such a tiny portion of the whole DNA, and indeed of the oocyte, that the problem of 'non-identity' is not likely to apply. If the 'non-identity' problem does not apply, then the 'equivalence thesis' applies. If Brassington is right to say that the mitochondrial DNA is such a tiny portion that it does not affect the identity of the child, then a child, *the same child*, is likely to be born, whether the portion of faulty mitochondria in the egg has been substituted or not. Therefore, it would be true to say that *a child* (*the same child*) will be brought into the world, either suffering from mitochondrial disease or clear of mitochondrial disease. It would not be true to say in this case that *one* child would be brought into the world who suffers from mitochondrial disease, or *another child* would be brought into the world clear of mitochondrial disease (as it would be if one embryo rather than another had been implanted). Thus the 'equivalence thesis' is still relevant here, but Brassington leaves the reader to make up their mind about this thorny issue.

The interest of applied ethics and bioethics in science has traditionally been raised mainly, though not exclusively, by scientific research that impacts upon human health – genetic and genomic research, for example, regenerative medicine research, embryonic stem cell research, and research that uses humans or other animals as research subjects. In this collection we wanted to cover science more widely, at the risk of sacrificing perhaps some depth to that purpose, and the rest of Part I of this collection is devoted to areas of scientific research which seem less directly involved in the protection or promotion of human health, but which, instead, as the authors show, are likely to have a profound impact on human health and welfare.

Catherine Rhodes discusses scientific research in relation to biosecurity (Chapter 6). Her chapter is (sadly) very timely, given the increasing global threats of attacks involving pathogens. She stresses that scientific freedoms are to be exercised within the context of certain responsibilities, which in some cases justify constraints on those freedoms. Responsibility to prevent certain threats from materialising falls on a large pool of actors, she argues: scientists, but also journal editors, scientific academies and national and international policy groups. Scientific research on pathogens is afflicted inherently by a profound tension: scientific work on pathogens yields huge public health benefits, but the same public health protection calls for a restriction on such work, or at the very least tight control of the release of information. International and national policy, Rhodes points out, increasingly hold scientists responsible for public health protection; but, as she notes, there must be recognition of reciprocal responsibilities of scientists and policymakers, to develop effective international policies that can mitigate the tension inherent in this area of scientific research.

David Lawrence in Chapter 7 focuses on robotics and artificial intelligence. After the Industrial Revolution of the 1800s, we are now used to the idea that machines carry out tasks traditionally performed by humans. Experimental robots are extremely impressive devices. We know already that they are widely applied to surgery, but, as Lawrence explains, it is possible now to emulate proprioception, tactility, visual processing, object recognition, walking and running. As with other areas of scientific research, the results are received either with enthusiasm or with worry. Automation is perceived by some as a threat; many human professions, it is feared, will disappear, being replaced by 'better robotic versions of ourselves'. The work market will be steered to make space for electronic engineers and similar professionals, to the detriment of the variety of what individuals, with their own unique talents and skills, can offer to society. The prospect of the development of conscious, thinking machines is even more disquieting. Artificial intelligence challenges regulation and policy around liability, ownership, employment and more. But Lawrence points out another preoccupation, which links his chapter to my earlier chapter. Scientific development challenges what it is to be human. I posed the question of how much of 'us' can be substituted by, say, robotic parts, before 'us humans' become something else. Lawrence poses the corresponding question: how much should an artificial life, a robot, understand, feel or think before it is regarded as 'one of us'? This metaphysical question raises a number of moral questions relating to how beings ought to be treated, moral questions about where on the ladder of moral status a being should be before it is entitled to equal concern and respect.

Part I of the volume concludes with Chapter 8 by Lucio Piccirillo. Piccirillo offers some reflections on the importance of science and freedom of scientific research, considering some of the important discoveries in the field of physics. In this chapter he takes on a role which, he argues, many scientists should take on in society, that is, to explain in plain and accessible language what their job is. He focuses on the Large Hadron Collider as an example of big science and on the Markov chain as an example of small science; first he explains what these are and what their purposes are, and then he offers some reflections on freedom to pursue both big and small scientific projects. He makes two seemingly straightforward points: the first is that insofar as science can yield advantages for humankind, science is a prima facie good, and thus scientists should enjoy a significant degree of freedom. The second is that in order to obtain this degree of freedom, society at large must receive clear and accessible information about the purpose and methods utilised in various scientific disciplines. So far so good, but as Piccirillo himself recognises, both arguments raise a number of complex issues. On the first point, Piccirillo distinguishes two types of impediments: financial constraints on the one hand, and ideologies and fears on the other. He discounts the latter as inherently detrimental to science, but accepts constraints based on resource rationing. But on the vexed problem of how scarce resources should be allocated, Piccirillo does not offer a solution or a method to begin to frame possible solutions.

On the second point, there are also questions to be asked. One is again about resources, and perhaps responsibilities. Is it really the job of scientists to communicate with the general public? Or is it the job of science correspondents, who often work as a liaison between scientists, on the one hand, and the public, through the means of the media? Corbellini, as we will see in Part II of this collection, raises doubts about whether it is scientists who ought to or are even best equipped to bridge the gap between science and society. Rather, the problem of the gap between science and society should be dealt with upstream, so to speak, at the level of public education in scientific disciplines and scientific methods. The public needs to be prepared to welcome and evaluate critically scientific discourses, and this cannot result from individual conversations of scientists, as clear as they might be, on specific issues. It is the state that has a responsibility to educate citizens, particularly during the phase of compulsory education, in scientific methods especially, in order to foster the analytic thinking that equips citizens to evaluate rationally facts, claims and scientific developments. Only this can protect us from various serious dangers, particularly from the exploitation and false claims divulged in the guise of scientific truths for fame or financial rewards – think of homeopathy or untested stem cell therapies, with no proven benefit or scientific credibility, which have been and are administered, generating lucrative payments, to patients affected by various serious diseases (*Nature* 2015); or think of claims around the ineffectiveness or harmfulness of vaccinations. There is little that can be done to convince someone to change their belief, say, that genetically modified foods destroy natural equilibrium, or to persuade someone who believes that vaccinations are harmful to vaccinate their children. Normally people do not abandon their strong beliefs, even in the face of evidence that they are likely to be mistaken. They often instead manipulate the facts to reconcile the cognitive dissonance, that is, to reconcile the contradiction in their minds. Therefore, a one-to-one or one-to-many conversation on a specific issue (say, a scientist explaining the science of genetically modified organisms) will be able, typically at least, to convey the message only to those already open to ideas that may contradict their beliefs or predispositions to certain beliefs. But being open to ideas that may contradict our predispositions is a complex and sophisticated skill to be acquired and developed. Thus, the conversation between scientists and the public at large must be preceded by a certain degree of formal education in science and in its methods, so that people can develop the critical skills that may enable them to approach scientific developments rationally and critically and to form more reasoned or rational beliefs.

Many of the contributors here, as mentioned earlier, participate in an ongoing international forum on freedom of scientific research begun in 2006, and have participated in academic and political debate both before and since. So, with this volume we want to contribute to an ongoing international bioethical and political debate on the ethics and politics of scientific freedom. We do not offer a collection from academics who have ideological affinity with us: we want to provide a truly multidisciplinary and open perspective on scientific developments in society and a critical reflection on the regulation of science. We hope to provide a balanced but progressive collection, which will promote further reasoned debate on the gap between science and society and on how to correct it.

References

Arendt, H. (1993), 'What is freedom?', in H. Arendt, *Between Past and Future: Eight Exercises in Political Thought*, New York: Penguin.

Berlin, I. (1982), *Four Essays on Liberty*, Oxford: Oxford University Press.

Bortolotti, L. (2013), 'The relative importance of uncomfortable truths', *Medicine, Health Care, and Philosophy*, 16.4: 683–90.

Genetic Alliance (2016), www.geneticalliance.org.uk/information/services-and-testing/predictive-testing (last accessed 26 September 2017).

Harris, J., and Keywood, K. (2001), 'Ignorance, information and autonomy', *Theoretical Medicine and Bioethics*, 22.5: 415–36.

Hobbes, T. (1651), *De Cive (The Citizen)*, www.constitution.org/th/decive.htm (last accessed 25 September 2017).

Mele, A. (1995), *Autonomous Agents: From Self-control to Autonomy*, Oxford: Oxford University Press.

Mill, J. S. (1989 [1859]), *On Liberty*, New York: Cambridge University Press.

Nature (2015), 'When right beats might. The final act in a long-running Italian saga should bring tighter controls on unproven stem-cell therapies, both at home and abroad', 24 February, www.nature.com/news/when-right-beats-might-1.16973 (last accessed 25 September 2017).

Palmer, J. (2013), *Parmenides and Presocratic Philosophy*, Oxford: Oxford University Press.

Takala, T. (2001), 'Genetic ignorance and reasonable paternalism', *Theoretical Medicine*, 22.5: 485–91.

G. Vattimo, and L. Cavalli Sforza (2006), 'Scienza o filosofia?', *Micromega*, 1: 7–24.

1

The influence of infection on society

Peter Lachmann

The main theme of this chapter is the enduring and extensive influence that combating infection has had on human life and society. This is a topic much neglected in accounts of human history. Moreover, the influence of infection is not restricted to humans but can be seen throughout the living world from bacteria and fungi to plants and animals.

The bacteriophages that infect bacteria have been invaluable tools to study molecular biology though their promise as antibacterial agents in medicine has not so far been fulfilled. The devastating effect of infection on the tree population in this country has been demonstrated by Dutch elm disease, and more recently by ash dieback, which have had a large effect on the overall tree population.

In animals, there is a very interesting review by Hamilton et al. (1990) who analysed why it was that primitive animals always adopt sexual reproduction as opposed to vegetative reproduction as used in many plants. They came to the conclusion that the advantage of sexual reproduction is that it provides a mechanism to reassort the genes that are concerned with resistance to infection at each generation. In other words, the reason we have sex is to combat infection. Here, however, I will restrict myself to discussing infectious disease and its effects on human societies.

It is likely that humans became significantly more susceptible to infectious disease as a result of the agricultural revolution about 10,000 years ago. This is less than 10 per cent of the period in which modern humans, *Homo sapiens*, have existed. For the first 90 per cent of human existence the communities were small, they tended to move about and not occupy the same site for long, and they had no domestic animals. Although evidence on the incidence of infection before the agricultural revolution is sparse, it is highly plausible that it was less. The coming of larger communities living at fixed sites led to their contaminating their water supplies with their own faeces and promoted orofaecal spread of infection. The fact that they lived in larger communities will itself have helped to spread infections by the respiratory route but perhaps the most important feature was the

Table 1.1 Examples of human infectious diseases of animal origin

Disease	Microbe	Animal source	Date of crossover	Location
malaria	parasite	chimpanzee	*c.*8000 BCE	
smallpox	virus	ruminant?	>2000 BCE	
tuberculosis	mycobacterium	ruminant?	>1000 BCE	
typhus	rickettsia	rodent	430 BCE	Athens
			1492 CE	Spain
plague	bacterium	rodent	541 CE	Justinian Plague
			1347 CE	Black Death
			1665 CE	Great Plague of London
dengue	virus	monkey	*c.*1000 CE	
yellow fever	virus	monkey	1641 CE	
Spanish flu	virus	bird, pig	1918 CE	worldwide
AIDS/HIV-1	virus	chimpanzee	c.1931 CE	
AIDS/HIV-2	virus	monkey	20th century	

Modified from Weiss (2001: 960)

domestication of animals. It is quite remarkable how many of the common infectious diseases appear to have a zoonotic origin (Weiss 2001). Table 1.1 shows some examples.

The close association of humans with the animals that they kept for hunting or as pets or as sources of meat and milk or hides or wool is likely to have caused a huge increase in the burden of infectious disease. This led to a human mortality pattern shown in Figure 1.1 where more than half of those born were dead before the age of 5 and thereafter mortality was more or less continuous, so that 50 per cent of the remaining population were dead before the age of 40. It is nevertheless interesting that the modal age of adult death, i.e. the age at which the largest number of people died in old age has probably been around 70 since biblical times at least – so that it is not that humans were at that time less capable of surviving for their three score years and ten, but that infectious disease often prevented them from doing so.

Infection, mortality and religion

Humans are probably unique in being aware of their mortality from early life. This awareness of mortality combined with the mortality pattern shown in Figure 1.1 will certainly have contributed to views on life and its signifi-cance and to have promoted the idea of life after death, or of repeated reincarnation, as a way of coping with the loss of so many young children and the constant threat of death throughout life. This can be seen as

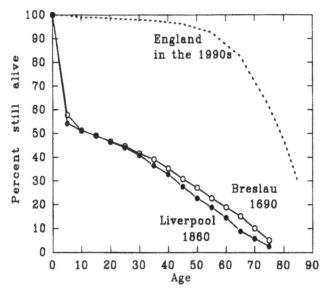

Figure 1.1 A history of mortality. Pattern of survival in seventeenth-century Breslau (which may have been typical for the times), nineteenth-century Liverpool (which had the lowest survival for any city in nineteenth-century England) and England in the 1990s (Cairns 1997)

contributing greatly to the growth of religions which place an emphasis on some continued existence after death, or on repeated incarnation.

It is, however, remarkable that there is one religion, the original teachings of Buddha, which regards repeated reincarnation not as an achievement to be desired but one from which one eventually wishes to escape. This escape is to nirvana, the state of absolute oblivion. This is achieved through enlightenment which requires freeing oneself of all desires. Buddha considered belief in god(s) as a desire, so that early Buddhism was anti-theistic. Buddha insisted that he was not a god. That humans achieve total oblivion after death is integral to modern secular beliefs except that it is no longer thought to be a necessity either to achieve enlightenment or to go through many cycles of reincarnation before oblivion can be achieved.

Making things better

However, the secularisation of society that accompanied the European Enlightenment really became widely established only after mortality patterns began to improve. This was brought about largely by three great contributions to fighting infectious disease.

The first improvement was better hygiene and public health, particularly the provision of clean water supplies and effective sewage disposal; but also more hygienic ways of preparing and storing food. An important early example showing the importance of clean water was the work of John Snow who in 1854 traced a large epidemic of cholera to the use of water from a pump in Broad Street in London. He persuaded the local authority to remove the pump handle and the epidemic promptly came to an end. It still took a long time thereafter before systematic provision of clean water (and chlorination where necessary) became enforced even in England. The 'great stink' in central London in 1858, by afflicting the Houses of Parliament, led to the provision of an effective sewage disposal system being devised by Joseph Bazalgette and built between 1859 and 1875. This again had a major beneficial effect on the incidence of cholera in London. More recently, the introduction of domestic refrigerators in the early twentieth century and of domestic freezers in the 1940s made a large contribution to reducing food-borne infection – as well as greatly reducing stomach cancer by reducing the use of wood smoke (with its carcinogens) to preserve food.

The second improvement was the introduction of vaccination. This first reached the West in the form of variolation against smallpox in the eighteenth century. Vaccination against smallpox using cowpox was introduced at the very end of that century. Both of these were quite empirical procedures, and are described in more detail in the account of smallpox below. The first involved administering a tiny amount of pus from a smallpox lesion which gave rise to a (usually) mild and localised attack of the disease. The second used lymph from a cow suffering with cowpox, a related virus, it having been observed that milkmaids who had caught cowpox while milking cows did not thereafter catch smallpox.

Vaccination based on an appreciation that infectious disease was due to microorganisms does not really start until the nineteenth century and the seminal work of Pasteur and of Koch and their associates and the rise of immunology. Their effects are not really seen to any great extent until the beginning of the twentieth century and particularly since mass vaccination against many diseases has been introduced on an increasingly worldwide basis (Table 1.2).

The discovery of antibiotics was originally made by René Dubos in 1938 who recognised that bacteria survive in soil by secreting substances that inhibit the growth of other bacteria. His original antibiotics, gramicidin and tyrocidin are still used topically but are too toxic for systemic use. Probably for this reason, he was, unjustly in my view, not given a Nobel prize. It was after the isolation by Florey and Chain in 1941 of penicillin, an antibiotic made by a fungus which is not toxic in mammals, that antibiotic discovery developed into a mega-industry. This was the third major factor in reducing the mortality from infectious disease in the period after the Second World War. No new antibiotics using the Dubos model have been discovered for some years now and the growth of antibiotic resistance among bacteria is giving rise to major concerns that bacterial infection may once more become a serious problem for the human population. The

Table 1.2 The effectiveness of vaccines

	Max. cases pre-vaccine	Year of max. cases	Cases in 1995	% change
pertussis	265,269	1934	4315	−98.37
measles	894,134	1941	309	−99.97
mumps	152,209	1968	840	−99.45
congenital rubella syndr.	20,000+	1964/65	7	−99.96
polio	21,269	1952	0	−99.99
adverse events	0		10,594*	

* For all these vaccines plus H.Influenzae, diphtheria and tetanus (data for USA from CDC)

development of chemotherapy for viral infections is a field that is still expanding. While effective drugs now exist against some important viruses, HIV being a good example, there are many viruses, measles for example, for which no effective chemotherapy yet exists; and others like flu where the treatments are not particularly effective.

Older practices that fight infections

However, even in a much more distant past, one can see in the prescriptions of various religions, practices that will have led to increased resistance to infectious disease. This was certainly not the reason why such practices were introduced − at a time when the causes of communicable diseases were unknown. Examples here include the prohibition of cannibalism. There are probably multiple reasons why particular religions forbade cannibalism while others did not, but it became the case that communities which rejected cannibalism survived better. This was, at least in part, due to the spread of spongiform encephalopathies, diseases that in humans include kuru and Creutzfeldt–Jakob disease, which can be spread by eating human brain. It is therefore fairly clear that cannibalistic societies were at increased danger of succumbing to the spread of these lethal diseases. The Fore Tribe in New Guinea very nearly died out from kuru in the early twentieth century before the Australian government who then ruled New Guinea forbade the practice of ritual cannibalism.

It is likely that male circumcision persisted in some communities for an analogous reason. There is little doubt that the spread of various sexually spread diseases, not just HIV, is reduced by circumcision and it is likely that this practice helped to reduce the incidence of potentially lethal diseases like syphilis and hepatitis B infection, and prevented women becoming infertile as a result of non-lethal diseases such as gonorrhoea. These consequences again were clearly not the reason for the introduction of circumcision but the improved survival of communities practising it will have favoured its survival.

Historical pandemics and their effects on society

It is quite difficult to imagine now the scale of mortality that was produced by major pandemics of infectious disease. There must be many great pandemics of which we have no record but there are several which certainly had a major influence on the course of human history.

The Athenian plague between 430 and 426 BC during the Peloponnesian War, probably contributed greatly to the demise of the Athenian empire. The organism causing this plague is not certainly known. Typhus, smallpox, measles and flu have all been suggested. So has typhoid but that is implausible. It is estimated that 25 per cent of the city's population died (75,000–100,000 people) (Littman 2009).

The Antonine plague, between AD 165 and 168, is likely to have accelerated the fall of the Western Roman Empire. This is believed to have been a smallpox epidemic and to have killed 3.5–5 million people (Wikipedia 2017a).

Plague is due to a bacterium, *Yersinia pestis* commonly known as the plague bacillus, and typically gives rise to 'bubonic plague' which is spread by fleas from rats to humans. During pandemics it can also spread directly from human to human – giving rise to the even more deadly 'pneumonic plague'. Plague has been one of the greatest scourges to afflict humans.

The Justinian plague (Wikipedia 2017b), which decimated the population of both the Byzantine Empire and the Persian Empire in the sixth to seventh centuries from 542 to about 750, may have killed 100 million people worldwide. It is argued by Holland in his book *In the Shadow of the Sword* (2013) to have made the Arab conquests possible. The Arabs, in his view, lived in dispersed settlements much further south, somewhere near Yemen, where rats were much less frequent and so had not been exposed to decades of the Justinian plague. This allowed them with relative ease to conquer both large parts of the Byzantine Empire and of the Persian Empire. Holland suggests that Islam originated as a need to control these great new acquisitions and followed the Arab conquest rather than preceding it. He also believes that the original Mecca was far further south than Mecca is now and in consequence out of the range of the plague.

The Black Death was also due to *Yersinia pestis*. The Black Death in the early fifteenth century was a catastrophe on a scale that is now quite difficult to imagine. It is estimated that more than one-third of the population of much of Europe died over a relatively short period and in some communities in East Anglia it has recently been shown that the production of ceramics fell by 70 to 80 per cent reflecting the enormous loss of population. It is certainly regarded as having caused the end of serfdom since there was such a shortage of labour that labourers were no longer bound to stay at their original sites of serfdom but could sell their labour throughout the country. This so-called 'second wave' of plague continued for centuries in Europe with intermittent outbreaks culminating in Britain in the great London Plague in 1665. This also caused a very substantial death rate.

It has been speculated that the decline of plague after this time was the result of the gradual replacement of the indigenous black rat by the larger brown Norway rat which appears to be a less congenial host for the rat flea. Rather later, the population became less tolerant of carrying fleas too!

Plague, however, still persists. In parts of the United States it is endemic and there was a small outbreak in North Africa during the Second World War which, greatly exaggerated, forms the subject of Camus's novel *La Peste* (Camus 1947). However, *Yersinia pestis* is sensitive to antibiotics and this has caused plague largely to die out, although the emergence of antibiotic-resistant plague bacteria is now giving rise to concern.

Most other major plagues that we know of have been caused by viruses rather than bacteria. The most devastating of these in human history has been smallpox. However, it is so far the only human pandemic disease to have been eradicated from the planet (Glynn and Glynn 2004; Rhodes 2013). Smallpox has been a scourge among humans since ancient times. It is probably of zoonotic origin and related pox viruses occur in many mammalian species. It is known to have been in the Far East centuries ago and it is there that the first attempts at prevention were established by variolation. This was introduced from the East to Turkey where it was used particularly to immunise the Circassian women who were used as slaves by the Turks and whose freedom from smallpox scars was held to be of greater importance than the occasional death. From Turkey, variolation was introduced into England through the good offices, in part, of Lady Mary Wortley Montagu (as shown in her letter below) and became reasonably widespread.

Extract of a letter from Lady Mary Wortley Montagu in Constantinople to Mrs Sarah Chiswell, written in 1717 (Lynch n.d.: Letter XXXI)

The small-pox, so fatal, and so general amongst us, is here entirely harmless by the invention of *ingrafting*, which is the term they give it . . .

Every year thousands undergo this operation; and the French ambassador says pleasantly, that they take the small-pox here by way of a diversion, as they take the waters in other countries. There is no example of any one that has died in it; and you may believe I am very well satisfied on the safety of this experiment, since I intend to try it on my dear little son . . .

I am patriot enough to take pains to bring this useful invention into fashion in England; and I should not fail to write to some of our doctors very particularly about it, if I know any one of them that I thought had virtue enough to destroy such a considerable branch of their revenue for the good of mankind.

Variolation was never entirely safe and people who were variolated had to be isolated for some period to avoid infecting others. However, it was this practice that enabled Edward Jenner to do a trial of vaccination with cowpox in one patient. This was given to a boy, James Phipps, who was subsequently variolated and the variolation was shown not to take. This led to the widespread introduction of vaccination with cowpox and subsequently with vaccinia (whose exact animal origins are unknown) to prevent smallpox – the first great triumph of vaccination at the end of the eighteenth century.

Although smallpox epidemics occurred throughout the Old World, it was unknown in the New World and it was smallpox (together with measles) that enabled small numbers of Spanish conquistadores to conquer Mexico and Peru. Although other explanations, like more ruthless warfare, better guns, or horses have been quoted, it is almost certainly the case that they brought smallpox with them that caused their rapid success (Oldstone 2010). It is recorded that Cortez had a slave in his entourage who was incubating smallpox when they arrived in America. In populations with no herd immunity to smallpox, the mortality was extremely great. Similar findings were recorded in California where the establishment of mission stations was always followed by the death of much of the local population, the only survivors being the offspring of the native women and the soldiers guarding the missions (and probably the priests too). It is strange that the cause of this mortality was never really considered at the time.

Disgracefully, smallpox was also used in biological warfare, even by the British. Pox viruses are extremely stable in the environment and blankets that had been exposed to a smallpox patient were given to native Americans during the siege of Fort Pitt in 1763 to infect the natives – an entirely shameful episode and quite probably not unique.

The eradication of smallpox by a massive vaccination programme in the 1960s is one of the great triumphs of preventive medicine. It would be very difficult to achieve now because of the increased emphasis on individual consent. The smallpox eradication campaign was done with what is called 'community consent' so that the local authority would agree and then all the available children would be lined up and vaccinated. Attempts to eradicate measles and polio, two other great scourges of humankind, are still not fully complete although they are very slowly reaching completion.

Eradication by vaccination is, of course, possible only where humans are the only host for the organism and cannot be achieved if there are alternative hosts such as in the case of yellow fever and many other diseases.

Another major epidemic, of which there are a few human survivors who still remember it, was the flu epidemic that followed the First World War in 1918. It is reckoned that this killed some forty million people – more than were killed in the war. Flu is a constant companion of humans and has given rise to frequent epidemics as well as the regular presence of flu in the population, but it is only occasionally that disastrous pandemics such as those of 1918–19 arise. Flu virus has an unusual genome where there are eight separate stretches of genome in the virus, which can therefore reassort rather than recombining and this gives the virus a great capacity of rapidly induced changes in its antigenic structure. There has more recently been a major outbreak of bird flu, which is contagious and can be lethal to humans, but which fortunately so far has not shown the ability to spread readily from one human to another. This again, however, cannot be guaranteed to continue indefinitely although the major haemagglutinin antigen of the bird flu virus (H5) is not known ever to have caused pandemics in man.

Another current global viral pandemic is HIV. There are two HIV viruses – HIV1 and HIV2. Both are lentiviruses but they are distinct. HIV1 is endemic in chimpanzees in Central Africa and HIV2 in sooty mangabeys in West Africa. The spread of these viruses to humans probably occurred some decades ago but they became a major problem much more recently, from the 1970s. It is highly likely (but difficult to fully establish) that the cause of this accelerated spread was the introduction into Africa of hypodermic syringes and needles for vaccination and other medical purposes. At that time the danger posed by their misuse after their original use was not recognised and no efficient disposal procedure was put in place. The syringes and needles allowed materials from animals, monkeys in particular, to be injected as part of native African medicine rather than just smeared as was previously the custom. This is likely to have caused the catastrophic explosion in the amount of HIV present and facilitated sexual spread – both homosexual and heterosexual and allowed spread to the USA, Europe and the rest of the world. Although relatively effective antiviral drugs against HIV have been developed, there is still no effective vaccine against the virus, which is highly immunosuppressive and whose cellular host is the T-lymphocyte, the major mediator of antiviral immunity. If this account of the pandemic spread of HIV is correct, it is a particularly tragic example of how interventions made with the best possible intentions can have catastrophic side effects.

Other infectious diseases have exercised a major effect on human society not in the form of pandemics but because they are continuously present in various human populations and have been major causes of mortality: among the most important of these is tuberculosis caused by *Mycobacterium tuberculosis*. Mycobacteria are intracellular bacterial infections which are difficult to neutralise and are contained by cellular immunity which is however also responsible for much of the pathology produced by the infection. Tuberculosis has been a major cause of death for centuries and, as is known from nineteenth-century literature, was regarded at that time with the same apprehension as cancer is at the present time. There is an excellent account of tuberculosis given by René Dubos (1987), which concerns attempts to control it. Some degree of immunity can be produced by infection with attenuated organisms, the so-called Bacillus Calmette–Guérin (BCG), but this is more effective in northern than in southern climates, though it does largely prevent everywhere the more lethal forms of childhood tuberculosis – miliary tuberculosis and tuberculous meningitis – and for this reason it is well worth persevering with. However, effective control of tuberculosis was only achieved when antituberculous chemotherapy was produced – notably streptomycin – followed by isoniacid and polyaminosalicylic acid. The growth of multi-drug-resistant tuberculosis at the present time presents a very serious hazard to public health globally in the next decades.

In and around the tropics, malaria is a major endemic scourge. It is caused by a protozoan parasite – *Plasmodium* – of which there are several species, *Plasmodium falciparum* being the most important in man. It is spread by the bite of the anopheles mosquito. Malaria is an important cause of chronic disease in adults and of death in children. Although there are effective treatments, it is still a major cause of death and morbidity in lower-income countries around the tropics.

Effects of infection on human genetic variation

Malaria is the canonical example of a rather different way in which infection has impacted human society – by its selective effect on human genetic polymorphisms. Plasmodia infect red blood cells and their ability to do so efficiently is influenced by the haemoglobin the cells carry. There is a haemoglobin variant – sickle cell haemoglobin – which when there is a single copy of this gene gives some slight protection in young children against malaria. This gene has been selected for in areas were malaria was common. Where subjects are homozygous for the sickle gene they suffer from sickle cell anaemia – an unpleasant disease which requires repeated blood transfusion throughout life. This, and a variety of analogous haemoglobin variants, have become a real problem for countries like Sardinia and Cyprus where malaria has been eradicated and where treating these haemoglobinopathies takes up a considerable fraction of their health budget.

Attempts have been made to eliminate these haemoglobin alleles. The Roman Catholic Church has given its approval to the testing of people before marriage and forbidding the marriage of two carriers of the disease, i.e. two heterozygotes. This is not a great idea. It is difficult to enforce and therefore doesn't work well and in addition it is a solution that is 'dysgenic', i.e. it leads to the increase in the number of the harmful genes rather than a decrease. If heterozygotes are not allowed to marry each other, there will be no breed-out of the homozygotes who carry two copies of the gene and an increasing percentage of the population will become heterozygous. The 'eugenic' solution would be to let people breed at will but where two heterozygotes breed together, the foetus should be checked for the presence of the gene and aborted if homozygous. With modern medical technology, this no longer requires invasive biopsy of the early foetus because genetic analysis can be done on foetal cells that pass into the maternal circulation. By this non-invasive technique the homozygotes can be identified early in pregnancy and aborted at an early stage. This still gives problems to those religions, largely Roman Catholics, who believe that the embryo acquires full human status at the moment of conception. This, however, is a modern idea, propounded first by Pope Pius IX in 1869, and has no origins in more ancient teachings. It was probably an attempt to bring the teachings of the Church in line with what was then known of human embryology. In this respect it was entirely misplaced. Pius IX and his contemporaries did not realise that the majority of fertilised eggs to which they ascribe full moral status do not even implant in the uterus; and, since they have had no opportunity to sin, they would form the majority population of heaven (or limbo) – a concept that I think would be foreign to the thinking of even the most right-wing Catholics.

There has been much written on the moral status of the embryo, that by the late Gordon Dunstan (1984) being a good example. I think it is widely accepted, among philosophers at least, that full human status cannot be granted unless there is some form of sentience and intentionality. There is no possibility of sentience when an embryo has no central nervous system and no sense receptors. If it can neither hear, nor feel, nor smell, or see, then it is not capable of being sentient and is not appropriately regarded as having achieved full or even appreciable moral status.

It is clear that the removal of these haemoglobin variants that once were important in resisting malaria need to be eliminated if the health services in many of these countries are to survive, and this is an area where it really is important that some rationality is applied.

The human genetic locus that shows the most variability is known as the major histocompatibility complex (MHC). These antigens are responsible for binding peptides derived from pathogens and allowing T-lymphocytes to react with them, and because there are so many antigenic variants it is important that there is a great assortment of these antigens. This is the group of antigens in mammals that correspond to the ones that Hamilton was referring to when he said that we had sex because of infection and they remain a vital component of our resistance. They are one of many reasons why excessive inbreeding in human communities can cause their demise because there are insufficient variants, and one particular infection can wipe out a whole population. This is another example of where a religious prohibition has survival value that is unlikely to bear any relation to the reasons for which it was introduced.

Conclusion

The consequence of improved hygiene and vaccination, antimicrobial and antiviral therapy has led to great changes in the mortality curves (see Figure 1.1) so that now it is relatively unusual for humans, in the developed world, to die of natural causes before the age of 60 and mortality has become concentrated in old age. As already said, this has led to a very great change in attitude to mortality and is perhaps more important in the secularisation of the world than is commonly realised. This has also led to a huge population explosion, particularly because of the reduction in childhood mortality, which is relatively easy to achieve and where the adoption of effective contraception has failed to balance the reduced mortality. Most religious prescription, particularly in regard to reproductive practices, is still that of an endangered species with enormous emphasis being placed on the duty, as well as the right, to reproduce. With the changes in society that place more emphasis on personal autonomy, even those restrictions which previously applied to procreation outside marriage have been abandoned and this effect may have even been exaggerated.

It is entirely self-evident that unless effective steps are taken to stop the growth of the human population, then there is a serious threat to the survival of mankind even in the medium term. No amount of reduction in CO_2 output, although highly desirable to protect against global warming, or in the consumption of other products needed for human life, will prevent catastrophe without the control of human population. There is sadly, in many parts of the world, still no serious attempt to see this done.

On the other hand, reduction in human population may be brought about by other means. The rejection of vaccination on irrational grounds, the abuse of antibiotics, the increase in global travel and threat of biological warfare, are all dangers to the future of the human race from infectious disease, as is the possibility, looking more serious now than at any time since the 1960s, of nuclear war. It may be that the second half of the twentieth century will come to be seen as a golden period, certainly with regard to infectious disease. The possibility that we will be unable to control drug-resistant infections such as tuberculosis, or of other bacteria, is not at all unreal, though it is possible that science and human ingenuity may still win through.

What is totally clear is that the changes in human behaviour necessary to prevent disaster will need the participation of those who understand the roots and ethics of human behaviour as well as the underlying science.

References

Cairns, J. (1997), Matters of Life and Death: Perspectives on Public Health, Molecular Biology, Cancer, and the Prospects for the Human Race, Princeton, NJ: Princeton University Press.

Camus, A. (1947), La Peste, Paris: Gallimard.

Dubos, R. (1987), *The White Plague: Tuberculosis, Man, and Society*, New Brunswick, NJ: Rutgers University Press.

Dunstan, G. R. (1984), 'The moral status of the human embryo: a tradition recalled', *Journal of Medical Ethics*, 10.1: 38–44.

Glynn, I., and Glynn, J. (2004), *The Life and Death of Smallpox*, London: Profile Books.

Hamilton, W. D., Axelrod, R., and Tanese, R. (1990), 'Sexual reproduction as an adaptation to resist parasites (a review)', *Proceedings of the National Academy of Sciences of the United States of America*, 87.9: 3566–73

Holland, T. (2013), *In the Shadow of the Sword*, London: Little, Brown.

Littman, R. J. (2009), 'The plague of Athens: epidemiology and paleopathology', *Mount Sinai Journal of Medicine*, 76.5: 456–67.

Lynch, J. (ed.) (n.d.) *Letters of the Right Honourable Lady Mary Wortley Montagu; Written during Her Travels in Europe, Asia, and Africa, to Persons of Distinction, Men of Letters, etc. in Different Parts of Europe* (London, 1790), https://andromeda.rutgers.edu/~jlynch/Texts/montagu-letters.html (last accessed 26 October 2017).

Oldstone, M. B. A. (2010), *Viruses, Plagues and History: Past, Present and Future*, Oxford, Oxford University Press.

Rhodes, J. (2013), *The End of Plagues: The Global Battle Against Infectious Disease*, New York: Palgrave Macmillan.

Weiss, R. A. (2001), 'The Leeuwenhoek Lecture 2001. Animal origins of human infectious disease', *Philosophical Transactions of the Royal Society of London. Series B: Biological Sciences*, 356.1410: 957–77.

Wikipedia (2017a), 'Antonine Plague', https://en.wikipedia.org/wiki/Antonine_Plague (last accessed 26 October 2017).

Wikipedia (2017b), 'Plague of Justinian', https://en.wikipedia.org/wiki/Plague_of_Justinian (last accessed 26 October 2017).

2

Scientific progress and longevity: curse or blessing?

Simona Giordano

Background: life, death and the elixir of long life

Many myths and epics, from various eras and geographic locations, mirror the human yearning for longevity. The Asian Epic of Gilgamesh is one of the oldest (2600–2500 BC), but there are many others: the Garden of Eden (in the Bible), the Holy Grail, the Fountain of Youth, the Philosopher's stone, to mention just a few. Many 'precursors' of modern scientists actively sought to find medical means to defeat death (a notable example is the alchemists, particularly the Chinese alchemists in the fourth and third century BC). In some sense, it may be argued that much, if not all, scientific effort is ultimately meant to prolong human life, the life of the planet, to give humans and perhaps other living beings the best chance to live as well as possible for the longest time, hopefully forever. Many of those who defend scientific freedom do so in the name of human welfare, longevity and freedom from illness and disability (see Chapter 15 in this volume).

The longing for immortality is obviously the other face of the anxiety about death. The finitude of human existence is at the heart of Western philosophy across the centuries. As Lachmann (see Chapter 1 this volume) points out, the awareness of death is probably exclusive to humans among all the animals, and this has probably prompted the birth of theist religions that promise some kind of afterlife or reincarnation. Many philosophers and schools of philosophy since antiquity have also attempted to ease worries about death. The Eleats believed in an eternal reality, from which we come and towards which we are directed. Life is a passage in an uninterrupted journey. Later, in Plato and much of the Platonic tradition, death acquires a positive value; through death the soul will be finally freed from the 'cage' of the body. Plato, in his *Apology*, writes:

> Either to be dead is not to exist, to have no awareness at all, or it is, as the stories tell, a kind of alteration, a change of abode for the soul from this place to another. And if it is to have no awareness, like a sleep when the sleeper sees no dream, death would be a wonderful gain. (1984: 103)

Epicurus expanded on the idea that death represents the end of sensations, including good and bad:

> Accustom yourself to believe that death is nothing to us, for good and evil
> imply awareness, and death is the privation of all awareness; therefore a right
> understanding that death is nothing to us makes the mortality of life enjoyable,
> not by adding to life an unlimited time, but by taking away the yearning after
> immortality. For life has no terror; for those who thoroughly apprehend that
> there are no terrors for them in ceasing to live. Foolish, therefore, is the person
> who says that he fears death, not because it will pain when it comes, but
> because it pains in the prospect. Whatever causes no annoyance when it is
> present, causes only a groundless pain in the expectation. Death, therefore,
> the most awful of evils, is nothing to us, seeing that, when we are, death is
> not come, and, when death is come, we are not . . . The wise person does not
> deprecate life nor does he fear the cessation of life. (2017)

In spite of these (and other) efforts to calm uneasiness at the thought of dying, aversion to death is quite pronounced in humans and drives much of our individual and collective efforts. Contrary to what Epicurus may have hoped, humans have continued across the centuries to fight death and deadly diseases, and to extend and ameliorate their quality of life.

Science, including medical science, developed and changed drastically during the Renaissance: Kepler, Copernicus, Newton, Da Vinci, Vasalio, Descartes and many others revolutionised the understanding of the universe, of humankind and of other animal kingdoms. The changed landscape paved the way for the great discoveries and innovations of the second part of the 1800s and of the 1900s.

Science has had a significant impact upon longevity: Lachmann highlights that the main determinant in the reduction of mortality since the 1900s has probably been the development of epidemiological science, which has allowed control of infectious diseases (see Chapter 1 in this volume). There is wide agreement among demographers as to the factors that have extended human life: tackling infant mortality, particularly through the mastery of infectious diseases, was one of the most important in the second part of the 1800s. During the following century, biomedical sciences also tackled much better the diseases of older people, particularly cardiovascular diseases. Improvement in living conditions, provision of clean water and sewage disposal, better nutrition, better education, better income and better medical care have all contributed to doubling life expectancy in just over a century, initially more prominently in middle- and high-income countries (Bonicelli and Sciarretta 2005), and now across the globe (WHO 2015a). Among all these factors, it seems that the contribution of science and medicine, through immunisation, vector control and provision of drugs, is probably the most substantial (Gratton and Scott 2016: 19).

Moreover, cellular and genetic therapies offer the possibility of treating or preventing many serious diseases and extending human life further.

Transplantations, now seen as established treatment, were only developed in the 1960s, and give us the opportunity to substitute old or damaged body parts with younger and healthier versions; genome editing revives the prospect of xenotransplantations. Cellular therapies can regenerate tissues, and if genetic and cellular research realised its potential, many degenerative diseases could be cured or even eradicated from humankind, with the result that we all could potentially live longer and disease-free lives.

There is further ground for optimism for the non-Epicurean. At the end of the 1990s it was discovered that lifespan depends on the bottom part of the chromosome (called the telomere), which protects the genetic material (Bodnar et al. 1998). Every time the genetic material is replicated during cellular replication, the telomeres become shorter, and thus they lose part of the genetic information they carry. Ageing is determined by this process, and this process ends when the cell has lost all the genetic material, cannot replicate itself and dies. A gene is responsible for the initial length of the telomeres and for the rhythm of shortening. Therefore, in principle, intervening on the length of the telomeres either with drugs or by genomic therapy could extend human life, in principle forever (accidents and illnesses aside). The hypothesis that human life could be extended indefinitely is contentious: a 2016 article published in *Nature* denies that this could ever be possible (Dong et al. 2016), but experiments on non-human animals show that this type of intervention may reverse the process of ageing (Broccoli et al. 1997).

Immortality may not yet be on the menu for us, at least not on this earth, but we live longer than the previous generation, and the next generations will live longer than us. This offers us hope for a long life, and is perhaps the fulfilment of one of our most ancient and rooted dreams. But, as we all know, the sweetest of all dreams can easily turn into the worst nightmare.

Introduction

Today men live on average 76 years and women 83, and average lifespan extends itself by about three months every year. According to some studies people will soon live on average 120 years, and adolescents today have a life expectancy of 100 years (Bonicelli and Sciarretta 2005). In the UK, by 2040 one in seven people will be aged over 75 (GOS 2016). A real demographic revolution is thus well under way. This raises a number of questions and concerns. Some are metaphysical in nature: longevity raises questions about what it is to be human. If finitude is an inherent feature of humanity, how long can our life be before we become *something else*?

Biomedical sciences also raise metaphysical issues. Think of transplantation: naturally we know that nearly all our cells are replaced regularly, so in one sense it shouldn't worry us that bigger body parts are replaced. But how many parts of 'me' need to be substituted before 'I' become 'someone else'? Are there some specific organs that are constitutive of personal identity?

The heart, or the brain, or something else, or nothing? Would I still be me in some significant sense, if, for example, my body parts (some or all) were replaced regularly, every 50 or 60 years, say? And how about replacement with other animals' body parts, or with artificial parts? What is 'I' a function of? Of the way I look? The face I have? The skills I have? The materials I am made of? The memories I hold? Would I still be me if I had, say, a brain replacement, or if my vital organs were not made of organic material? Or would I be someone else? (Bonicelli and Sciarretta 2005).

I will not discuss in this chapter the metaphysical issues relating to what it is that makes 'you' and 'I' respectively *you and I*, and what it is that constitutes the essence of personal identity. One of the reasons for leaving the metaphysical questions aside is that in the debate about ageing and longevity, these metaphysical issues are sometimes used simply to contradict the value of science. For example, it is sometimes argued that to strive for longevity and immortality is to deny *the dignity* of humankind as it is (Ramsey 2009). The appeal to the metaphysical issue of what a 'human' is thus is used ad hoc in a non-elaborated attempt to discredit certain practices.

Another similar objection to science, which I will not consider here in any depth, is the one that argues that the attempt of biomedical sciences to eradicate diseases is an attack on the dignity and value of people who have those diseases. This is a particularly common argument with regard to genetic testing, particularly prenatal genetic testing, genome editing and mitochondrial DNA replacement. For example, some see prenatal genetic testing for chromosomal disorders as an attack on the dignity of people born with Down's syndrome – and similar arguments apply to other conditions as well (BBC 2016). An in-depth response to these concerns would require a separate analysis, but it is sufficient to note here that there is no reason why achieving a longer lifespan should imply the ascribing of less dignity to those, or those generations, who have not been lucky in the same way; similarly, attempting to eliminate certain diseases, or supporting science to do so, in no way involves undermining the dignity of people who are born with or have developed those diseases. One can consistently believe that it is not immoral (or that it is indeed good) to attempt to cure or even eliminate cancer, dementia, spinal cord diseases and many others, while recognising the equal dignity and value of people who suffer from those conditions.

I will thus leave aside the metaphysical issues relating to personal identity, or arguments suggesting that science, particularly biomedical sciences, represent an attack on human dignity. I will instead focus on the practical and ethical concerns surrounding longevity: how is society going to cope with an increasing number of long-lived people? Greater numbers of older people in society may mean an increase in dependency, infirmity, dementia epidemics, medical spending. Will the workforce cope with the increasing demands of the older sections of the population?

The implications of population ageing are vast: they spread to housing, pensions, family life, family responsibilities, transport, education, working patterns and healthcare systems (GOS 2016). The 'sweetest of all dreams' can easily mutate into a looming crisis. The worries are wide-ranging: levels of ill health and disability will increase, the workforce will become increasingly reduced, and chronic conditions, multiple morbidities and cognitive impairments will become more common, raising long-term expenditure to unknown levels. At the same time families will face increasing pressure to balance care with other responsibilities, particularly work. As the population ages, so will the workforce: how can the nation's economic well-being be preserved? (GOS 2016).

I suggest that the demographic changes are inevitable and irreversible; but they are not a curse: they are to be welcomed as one of the greatest triumphs of humankind (WHO 2002a: 6). Many worries surrounding longevity (of individuals and of our species) result from a misconception of old age as a season of dependency and burden, on persisting stereotypes of the old as frail and useless, and on misunderstandings relating to disease and old age. Moreover, many important steps can be taken to prevent certain negative outcomes from materialising.

The demographic revolution: facts and myths

In the last century the world has faced a real demographic revolution. 'In 2010, an estimated 524 million people were aged 65 or older – 8 per cent of the world's population. By 2050, this number is expected to nearly triple to about 1.5 billion, representing 16 per cent of the world's population' (WHO 2011). The main drivers of this ageing population are an increase in life expectancy and a decline in fertility and birth rates. Estimates for 2014 predicted that over the entire world, the number of over-sixties will double from approximately 11 per cent to 22 per cent between 2000 and 2050 (WHO 2014). Already in 2015, it appeared that people over 60 represented over 30 per cent of the population in Japan, and this proportion is predicted to grow over 30 per cent in many more areas of the world (WHO 2015a).

> The absolute number of people aged 60 years and over is expected to increase from 605 million to 2 billion over the same period . . . The number of people aged 80 years or older will have almost quadrupled between 2000 and 2050 to 395 million. There is no historical precedent. (WHO 2014)

The *fastest* increases are predicted to occur in low- and middle-income countries. For example, in China the number of those over the age of 65 is likely to increase to 330 million by 2050 from 110 million in 2011, and by then there could be 100 million people in China over the age of 80 (WHO 2011). For countries in which the demographic changes occur quickly, giving less time for adjustment, the strain on national infrastructures, especially the national healthcare systems, is likely to be significant.

Ageing populations are at the centre of debates on social and economic life. As Weisstub puts it: 'Longevity coupled with economic reality is a frightening cocktail for societies to bear' (2015: 150). Phillipson adds:

> Concerns about the most appropriate way of resourcing such populations, their impact on standards of living, and relations between age groups and generations feature prominently in public debate and discussion. The 21st century will without question be a time when all societies take stock of the long-term impact of demographic change and the implications for managing and organizing a major area of social and economic activity. (2015: 80)

Western governments have considered population ageing as 'a mixed blessing' (Phillipson 2015).

Longevity raises a number of ethical, social and political issues as well as issues of global justice. Will the labour force be able to assist the ever-growing proportion of older people? Will healthcare systems be able to cope with the demands of a long-lived generation? It also raises ethical and political issues of intergenerational and global justice: what do we owe to future generations? And what do we owe to each other globally? Until 2000, the European Union did not express particular concern over the ageing population. For example, in the OECD document *The Welfare State in Crisis* (cited in WHO 2000), the issue of long-term care was given only marginal coverage. Later, however, especially in the last decade, the WHO has been increasingly worried about demographic changes and the ability of states to cope.

However, a few points need to be clarified. Many older people continue to work in either the formal or informal labour sectors (WHO 2002a; 2012). Moreover, the productivity of the older person, overall, does not seem to be lower than that of younger workers; in fact, according to some studies, older workers are more efficient than younger workers (Russo et al. 2006; Heidemeier and Moser 2009; Staudinger and Bowen 2011; Backes-Gellner and Veen 2013). Experience, knowledge and insight may compensate for some of the losses that may accompany ageing (these and others will be discussed later).

It should also be noted that people after retirement age often take responsibility for household management and childcare, which allows younger adults to work outside the home. In this way, they contribute actively to the labour market. Older people often offer the skills and experience accumulated during their working life in the voluntary sector, acting as volunteers in schools, communities, religious institutions, business and health and political organisations (WHO 2002a; 2012). Many long-lived people, then, far from being a 'burden' on society, provide an important contribution to the fabric of society, and ultimately benefit the overall economy. However, it is a contribution that, not being always directly remunerated and thus not being directly a part of the complex financial system of paid labour, tends to be overlooked, and its value therefore tends to be underestimated.

There is another aspect that is noteworthy: Gratton and Scott (2016) propose a number of solutions to the issue of sustainability. They note that the three-stage life (education/working life/retirement) was relevant when average life expectancy was around 70 years, and is no longer relevant when people can expect to live well beyond 100 years. Therefore, the structure of life, and working life, needs changing. We will all need to work much longer, and this seems inescapable; but working patterns will also need to change – for example we will have to retrain later in life, as the skills acquired during our earlier years may be obsolete during our seventies and eighties, and it is likely that the working environment will also become more flexible. Interestingly in this context, Gratton and Scott note that science will be one of the sectors that are likely to obtain an increment as the population ages:

> Greater numbers of older people will create a demand effect to which sectors and market prices will respond. So, for example, it is likely that medical research focused on longevity and bioengineering will be significant growth sectors and the service sector will shift towards healthcare and service provision.
>
> Environmental concerns and sustainability will also exert a substantial impact on prices and resources and the relative size of different sectors. We are on the cusp of substantial shifts in energy provision and, if energy scarcity continues and energy prices rise, then there will be significant innovations in energy creation and resource conservation. The same is true of food supply, where there is an expectation of radical innovation especially in combination with genetic engineering and health concerns. (2016: 50–1)

The next section will look at the issue of healthcare needs, because one specific concern relates to the ability of the healthcare system to cope with the demands of an ageing population (OECD 2006). I will evaluate the changes in disease patterns that appear with increased longevity, and I will argue that these changes need to be understood. We will see that several diseases typically associated with old age are not an inevitable consequence of age, and there is a lot that can be done by individuals and collectively in order to reduce their incidence.

Longevity and healthcare

Becoming older, as is well known, exposes us to some ailments. The greater the number of older people, one may believe, the greater the demands on healthcare systems. However, whereas it is undeniable that older people may have greater healthcare needs than younger people, the relationship between longevity and *healthcare requests* or *healthcare access* is not linear.

First, not all older people are unhealthy, and not all young people are healthy; health needs are highly variable within individuals, and not just within age groups. But even if it were true that older people (say the group over 60) have greater healthcare needs than younger people, it does not follow that they actually *demand or access* the healthcare system; their needs thus do not automatically translate into spending. For example, in low- and middle-income countries the increase in healthcare needs does not result in higher demand on the healthcare system, due to a number of factors, including barriers to access (WHO 2015a). Also, in high-income countries, research shows that those with chronic conditions tend to use more healthcare than those who do not have these conditions. But among those with chronic conditions, people with additional functional limitations use healthcare services more than any other group. So, there are variations among individuals *and within groups*. Moreover, even in high-income countries, where there may not be particular barriers to access to healthcare services, people with lower socio-economic status tend to access healthcare less than other groups, regardless of their needs (Alecxih et al. 2010; Terraneo 2015).

It is thus extremely difficult to predict the impact that population ageing has or will have on healthcare expenditure. Even if barriers to access and social inequalities were to be eliminated, so that healthcare access and demands matched healthcare needs, the link between longevity and healthcare expenditure is not linear (WHO 2015a).

Recent research indicates that in high-income countries the peak of healthcare demands is around the age of 65–70; after that time demands decrease (Oliver et al. 2014; Kingsley 2015). Historical analyses also suggest that ageing may have less influence on healthcare expenditure than other factors. Research conducted in the US between 1940 and 1990 found that ageing

> contributed to only around 2% of the increase in health expenditures observed during the period. In comparison, technology-related changes in practice were responsible for between 38% and 65% of growth, increasing prices were responsible for between 11% and 22%, and growth in personal income was responsible for between 5% and 23%. Similarly, research on expenditures in France between 1992 and 2000 found the contribution of ageing to be relatively small, with the impact of changes in clinical practice being almost four times as large. (WHO 2015a: 96)

Therefore, the claim that population ageing will result in increased healthcare expenditure is simplistic.

A related worry is that population ageing is correlated with changes in disease patterns, and this in itself poses novel challenges for healthcare systems. While this is true to an extent, the relationship between these changes and longevity needs to be understood.

Disease patterns and longevity

In a 1998 document, the World Health Organization (WHO) reported that changes in the population structure affect disease patterns (WHO 1998a). The WHO lamented that what it called *non-communicable diseases* (NCDs) have become the leading causes of death both in 'industrialized countries' (WHO 1998a: 14) and 'developing countries' (WHO 2017). The main NCDs listed in 1998 were (WHO 1998a: 14):

- cardiovascular diseases;
- hypertension;
- stroke;
- diabetes;
- cancer;
- chronic obstructive pulmonary disease;
- musculoskeletal conditions (such as arthritis and osteoporosis);
- mental health conditions (mostly dementia and depression);
- blindness and visual impairment.

In 2002 the WHO again reported that NCDs may be significant and costly causes of disability and reduced quality of life (WHO 2002a: 34), and can be expensive to treat and long-lasting (WHO 2002b).

More recent research, however, amends the picture significantly:

- NCDs kill more than 36 million people each year.
- Nearly 80 per cent of NCD deaths – 29 million – occur in low- and middle-income countries.
- More than nine million of all deaths attributed to NCDs occur before the age of 60; 90 per cent of these 'premature' deaths occur in low- and middle-income countries.
- Cardiovascular diseases account for most NCD deaths, or 17.3 million people annually, followed by cancers (7.6 million), respiratory diseases (4.2 million) and diabetes (1.3 million).
- These four groups of diseases account for around 80 per cent of all NCD deaths.
- They share four risk factors: tobacco use, physical inactivity, the harmful use of alcohol and unhealthy diets (WHO 2017).

This research shows that NCDs are *not* diseases of the elderly and are not a result of longevity. They may afflict all sections of the population, regardless of age. In higher-income countries they are more often associated with old age, either because people tend to become affected later, or because people live for longer periods with their disease, due perhaps to better healthcare and better living conditions (Kalisch et al. 1998: §7.1). Moreover, NCDs are to a large extent preventable, and many prevention measures are highly cost-effective. Thus, again longevity and population ageing will not necessarily result in an increase in healthcare demands and expenditure (WHO 2002a; 2010).

As mentioned earlier, health and illness are highly subjective states, and it is simplistic to associate longevity with disease. However, it could be objected that as we age, we are subjected inevitably to a certain degree of molecular and cellular damage. This damage may result in functional impairments, which, in turn, may cause psychosocial afflictions (WHO 2015a). While this is true, it is also to be recognised that how susceptible individuals are to those losses, and how they respond to them, is subjective, and much can be done to prevent these losses, or to prevent a deterioration in people's quality of life once they have taken place.

We have seen that there are various factors that are responsible for the onset of NCDs; these diseases are not age related, and are much more directly correlated to behavioural, psychological, social and environmental factors than to chronological age. Tobacco smoking, for example, increases the risk of stroke and lung cancer, accelerates the decline of bone mineral density, muscular strength and respiratory capacity, and so may lead to important losses of functional capacity. Excessive alcohol consumption and a diet high in saturated fats and salt and low in vitamins and fibre are also associated with higher risk of cardiovascular diseases. From a psychosocial point of view, both decline in cognitive capacity and the sense of loneliness are often related, in the older person, to the loss of relatives and close friends. In their turn, lack of participation in activities and social isolation are related to a higher risk of disability, both physical and mental. These factors can be modified, and they do not automatically 'come with living longer'.

In what follows I will look specifically at the ailments that are often thought to be related to old age, and argue that the connection between these and longevity is not necessary: in fact living longer does not necessarily expose us to certain diseases. It is often other factors that do so, not primarily age. I will focus in particular on the relation between physical activity and NCDs. However, this should not be misinterpreted. What I point out does not gesture towards assigning individuals the sole or main responsibility for their own ill health. I simply point out that age is not primarily responsible for many NCDs, and that these are largely controllable and preventable. But whether or not individuals are in a position to make lifestyle changes or to exercise sufficiently for their health is not only a function of their conscious choices and will. Whether or not people are able to exercise depends on many factors (working patterns, family responsibility, costs, accessibility, information and infrastructures). Later in this chapter I will discuss some key actions across sectors that may reduce barriers to healthier longevity.

Musculoskeletal damage

A condition that afflicts older people more than other groups is musculoskeletal problems. There are various types of musculoskeletal problems and various degrees of impairment. However, it is well known that older people are, for example, subject to falls more than other age groups. Falls are a major cause of disability and even mortality due to injury (Dhital et al. 2010; Gillespie et al. 2012; Hoops et al. 2012; Lee et al. 2012; Karlsson et al. 2013). The fear of falling may also affect quality of life, providing a significant limitation on daily activities (Department of Health 2001).

The effects of falls depend to an important extent on bone mineral density (WHO 1998b). Bone mineral density decreases to some extent physiologically with age. Proportionally, the risk of bone fractures increases (Cheng et al. 1997). Loss of bone mineral density, also known as osteoporosis, begins at around the age of 40, and mainly afflicts women, especially after the onset of menopause. Hormonal changes (particularly decline in oestrogens) are partly responsible for such bone thinning, but decrease of muscle mass is also a major cause of bone mineral loss. Research has shown that after a peak in early adulthood, muscle mass declines with age (Rantanen et al. 2003; Cruz-Jentoft et al. 2010). Both the number and size of muscle cells decrease. Because muscle mass has a direct impact upon bone mineral density, weight-bearing exercise is a standard treatment for osteoporosis and osteopenia. Bone mineral loss thus can be slowed down significantly with exercise. Neuromuscular coordination, proprioception and postural stability also contribute to the prevention of falls (Dargent-Molina et al. 1996), and these can all be manipulated with regular physical activity (Bacher et al. 2002; WHO 2002a: 28; Chubak et al. 2006).

But even when someone *already* suffers from disabling loss of bone mineral density, with consequent reduced mobility and independence, exercise can be an effective treatment (Suominen 2006). Research provides slightly different data on the amounts and regimes of exercise required for older people already affected by different musculoskeletal conditions (Moreira et al. 2014). However, a positive correlation is observed between resistance training, bone mineral density and muscular strength. Some research shows that after only six months' strength training, muscular strength increases by 9 per cent (lower body) and 18 per cent (upper body) in people aged 70–79 (WHO 2002a: 5; Gianoudis et al. 2012; Ribeiro and Neri 2012).

Developing muscular strength is not just an effective prevention and treatment for osteoporosis: walking, climbing stairs and overall mobility rely on muscular strength (Studenski et al. 2011), particularly leg strength, and when this is poor, a person clearly becomes more dependent and increasingly frail. It could be argued that the quality of life of the long-lived person is thus to an important extent dependent on one single factor: *muscular strength*, particularly leg muscular strength, and this is highly controllable (HM Government 2014).

Cardiovascular diseases

It might be believed that living longer exposes us to cardiovascular diseases because the heart, like any other muscle, loses strength as we age. In reality, other factors relating to lifestyle represent a more significant risk than chronological age. It is common knowledge that smoking, a diet rich in saturated fats and salts, and a sedentary lifestyle are major risk factors for cardiovascular diseases. Recent research also notes a correlation between loneliness, social isolation and poor cardiovascular health (Courtin and Knapp 2015).

Epidemiological studies show that perhaps one of the most important factors for the prevention of premature death from cardiovascular diseases is physical activity, even in people with established heart diseases. Physical activity reduces the risk of death from cardiac disease by 20–25 per cent among people with established heart diseases (Shiroma and Lee 2010; Longobardi et al. 2012: S99). Positive changes are manifested in cardiovascular efficiency, blood lipids, blood pressure and thrombotic tendency (WHO 1998a: 6). Physical activity has a direct effect on the heart, as it may increase oxygen supply and improve myocardial contraction and electrical stability. Moreover, physical activity increases the diameter of the coronary arteries and this also contributes to reduction of blood pressure at rest (both systolic and diastolic). Blood lipid profile is also positively affected.

What is more, regular exercise prevents the occurrence of cardiovascular diseases *in those predisposed to them*. This is one of the clear cases in which genetics and biology can be, to an important extent, controlled through lifestyle. A person initially 'at risk' because of her family history can turn out to be a 'fitter than average' person. In short, diseases of the cardiovascular system, which, it is worth remembering, are the leading cause of death in many countries, are a result of many factors, and are not the inevitable consequence of longevity.

Reduced mobility

Mobility clearly has a direct influence on a person's functional capacity and quality of life. Loss of mobility means dependency, and dependency means costs. The first age-related changes that affect mobility are anthropometric changes. They are related to both stature and the joints' range of motion. People between 65 and 74 years old are approximately 3 per cent shorter than people between 18 and 24 years old. This is likely to result from the shortening of intervertebral disc spaces and resulting kyphosis. This shortening begins at the age of 30 for most people and increases after the age of 40.

The range of motion of the joints also declines physiologically. In addition, decrements in the sensory-motor system produce a decline in postural balance. Poor balance, posture and physical coordination may also increase the risk of injuries or falls due to false movements (Salzman 2010; Farley et al. 2011: 163).

Deterioration of the musculoskeletal system is not simply or mainly a function of age, but primarily of a sedentary lifestyle. Simply *moving* and working out in different areas (balance, agility, coordination and skeletal flexibility) reduces the spontaneous loss in these areas.

Metabolic diseases

Older people generally tend to 'put on weight'. It is thought that this is in part due to the decline in metabolic rate, that is, the ability of our body to burn calories at rest. Changes in metabolic rate depend on various factors, and age is not their primary cause. Metabolic rates depend to a very important degree on muscle mass. The more muscle mass we have, the faster our metabolism. This is why weight-loss training now no longer incorporates just aerobic and cardiovascular exercise, but also strength training. As we age, we experience a physiological loss of muscle mass. However, this physiological loss is exacerbated usually by the person becoming less active and can be counteracted by exercise.

Maintaining good metabolic fitness is important in order to prevent obesity, which is often associated with greater risk of coronary heart diseases and diabetes (Type 2). Energy metabolism can be improved significantly with exercise, and so again the reduction of metabolic rates can be delayed, prevented or contained.

Among the metabolic changes that may affect us, those relating to glucose metabolism are particularly worrisome. Poor glucose tolerance may lead to diabetes (Type 2, which generally occurs after the age of 40), which is characterised by variable degrees of insulin resistance and relative insulin deficiency. In its later stages, diabetes is associated with a number of serious disorders that have a great impact upon a person's quality of life (e.g. blindness, kidney diseases, heart diseases, stroke and peripheral vascular diseases severe enough to result in the amputation of a leg or foot). Again age is not the sole or even the main determinant of these metabolic changes. Genetic predisposition, but also obesity and physical inactivity, increase these risks. Whereas not much can be done to change our genetic predisposition, being genetically predisposed does not necessarily mean developing the disease: lifestyle can significantly alter the probability that people genetically predisposed to metabolic diseases will actually become affected by them. In particular, exercise improves the glucose metabolism, and so not only can prevent metabolic diseases, but can also assist in the treatment of those who have developed them (International Diabetes Federation 2013).

Cognitive functioning

Old age is often associated with decline in cognitive function, but this association must be understood properly. Research suggests that 'capacity to tackle complex tasks that require dividing or switching attention' (WHO 2015a: 55) may decrease as we age; also our ability to 'learn and master tasks that involve active manipulation, reorganization, integration or anticipation of various memory items' (WHO 2015a: 55) may decrease with age. However, the capacity to maintain concentration, avoid distraction, 'memory for factual information, knowledge of words and concepts, memory related to the personal past, and procedural memory . . . language features, such as comprehension, reading and vocabulary' (WHO 2015a: 55) do not deteriorate with age and remain stable throughout life.

Mental health: affective disorders and dementia

Affective disorders, particularly depression, are major afflictions for older people, but often, like dementia, go unnoticed at least during the early stages, as they are taken as an inevitable feature of ageing (Department of Health, UK 2001: 19). Again, however, the association between living longer and suffering from affective disorders needs to be properly understood. First, what is sometimes perceived as a feature of old age is in fact a disease that has multiple causes: these have often little to do with how long someone has lived (though they may be related to exposure to adverse life events, such as multiple losses of significant others) (Seitz et al. 2010). Second, the prevalence of depressive and anxiety disorders is in fact slightly lower among older adults than among younger adults (with the exception of older adults living in care homes) (WHO 2015a: 58). Third, whereas it is notoriously difficult to treat multifactorial diseases, it is possible to prevent them and ameliorate the conditions of those who suffer from them, regardless of their age.

With regard to dementia, the WHO predicts that the current number of sufferers (47 million people) is going to triple by 2050 (WHO 2015a: 59). In the UK, projections are for an increase in the overall number of cases from 822,000 in 2016 to 1.7 million by 2051 (GOS 2016: 77). But dementia is not simply a result of longevity. In fact, research shows that certain types of dementia may be prevented by reducing the risk factors for cardiovascular diseases.[1]

Once again, it appears that exercise has a major role in preventing and reducing the impact of mental illness among the long-lived. Earlier studies in gerontology suggest that regular physical activity helps to maintain and improve functional ability, health and mental well-being in the older person (Ruuskanen and Ruopilla 1995). More recent studies seem to confirm these findings (Weuve et al. 2004; Ikeda et al. 2012; Steinmo et al. 2014). It has been found that walking can help to prevent vascular-related dementia (HM Government 2014: 4, Annex A). It has also been noticed that people who perform regular aerobic exercise tend to suffer from depression less than inactive people, although it has been impossible to establish what the causal connection between the two is. Some studies suggest that exercise may make older people able to cope independently, and may thereby enhance self-esteem and confidence, which, in turn, may help to prevent or reduce depression (O'Connor et al. 1993). Physical exercise is also considered as a treatment for anxiety. It is associated with improved general satisfaction

---

Box 2.1 Physical (in)activity and spending

Physical inactivity costs the national health services nearly as much as smoking. Inactivity in the UK costs around £20bn per year (HM Government, 2014: 5).

Physically active people incur fewer direct medical costs than inactive people (Pratt et al. 2000; Wang et al. 2004; Hagberg and Lindholm 2006; Franklin 2008).

Physically active people have fewer periods in hospital, go to the doctor less frequently, and use less medication than inactive people.

People in work who are physically active have lower rates of sickness absence, fewer retirements on health grounds and are more productive (HM Government 2014: 12).

Absenteeism related to physical activity costs the economy 5.6bn per year (Cabinet Office for Urban Transport 2009, cited in HM Government 2014: 14).

Physical activity has indirect financial benefits: in London town centres in 2011 walkers spent £147 more per month than those travelling by car, thus contributing to the economy (Department for Transport 2012).

An increase in 1 per cent in physical activity could save 1.2bn over five years (HM Government 2014: 5).

Across a town of 150,000 people, if everyone walked an extra 10 minutes a day, 31 lives and 30 million£ per year would be saved (HM Government 2014: 4, Annex A).

People who are physically active also have 30 per cent–50 per cent reduced risk of getting colon cancer (Department of Health 2004), and approximately 20 per cent reduced risk of breast cancer (US Department of Health and Human Services 2008).

Public policy in England and in other countries now reflects and incorporates these findings.[2]

---

and well-being (WHO 1998b: 9.) Moreover, physical activity is associated with better social adjustment and cognitive functioning. It may enrich the social life of the elderly, as it may be a way of meeting other people. Social participation greatly affects the quality of our life, and isolation is associated with higher mortality among older people (Sugisawa et al. 1994).

Intergenerational relationships

As mentioned in the Introduction, there are various steps that can be taken to avoid certain negative outcomes. One element conducive to healthy ageing that has been identified in the literature is intergenerational dialogue and solidarity. Phillipson argues that part of the worry relating to longevity is the liberal assumption, embedded in many Western democracies, that the state has (through its welfare system) the main responsibility for vulnerable members of society. Increases in longevity and the related modification in population structure challenge this assumption. Phillipson argues that forms of solidarity need to accompany state responsibilities. He writes:

> [W]e need to think about new forms of solidarity both to replace existing institutions and to indicate the basis for alternative forms of social action. Four illustrations [can] be made to develop this point: first, identifying forms of cooperation that bring together different generational interests; second, reconnecting to the original vision of the welfare state; third, adopting a human rights perspective in old age; and fourth, restoring meaning and dignity to the end of life. (2015: 90–1)

Phillipson's suggestion echoes an older recommendation made by the WHO. In *Active Ageing* (2002a) the WHO recognised that negative attitudes towards older people result partly from a lack of interaction among age groups. In some societies, intergenerational dialogue is simply a part of life. In many Asian countries, for example, extended families live in multigenerational households. In many European countries, however, where the nuclear (and the blended) family has replaced the extended family, intergenerational dialogue has to be a conscious choice (WHO 2002a: 20).

Intergenerational cooperation would allow the older person to have some interests in common with younger generations, to exercise his or her skills (memory, learning and cognitive abilities) and to continue to contribute to the changing labour market. Moreover, intergenerational dialogue promotes the transmission of values between different age groups and a more positive and realistic attitude of the younger towards the older.

The WHO advised a number of key action points: schools and communities should provide intergenerational activities such as training in new technologies for older people and opportunities for lifelong learning (2002a: 52); intergenerational solidarity could be enhanced by 'supporting traditional societies and community groups run by older people, voluntarism, neighbourhood helping, peer monitoring and visiting' (WHO 2002a: 28).

Phillipson, similarly, offers some examples of existing intergenerational cooperation and provides ideas for further development of intergenerational solidarity:

> [An] important area for informal education has been the development of 'intergenerational learning' – that is, educational programs that link older with younger learners. Newman and Hatton-Yeo (2008) cite the example of the NUGRAN program at the University of Valencia, which creates learning experiences that cross the generations, involving older and younger adults together, with the aim of promoting greater contact, trust, and more positive

attitudes between them. The program began with 71 students in 1999 and had expanded to 1,000 by 2007. Achenbaum (2005: 61) provides similar examples from the United States, citing in particular the partnership between the University of North Carolina at Asheville and the North Carolina Center for Creative Retirement: 'Participants take classes that they design, conduct inter-generational programs to develop leadership skills, and analyze problems in the community and at the state level.' (2015: 90–1)

There are also, Phillipson points out, common interests:

> Pillemer et al. (2010) argue that older people represent an important source for creating solutions to environmental problems through volunteering and civic engagement, drawing upon their own knowledge and experience . . . Steinig and Butts (2010), discussing the U.S. experience, highlight the fact that intergenerational strategies can have a positive impact on the environment through shared sites and housing developments that bring generations together. (2015: 92)

There are many other key actions that can and perhaps ought to be implemented, in order to ensure the sustainability of a long-lived world. In this short chapter it is not possible to cover them in detail. Gratton and Scott have provided an accurate analysis of the various comprehensive changes that need to be taken in order to ensure that longevity is not a curse, but an exciting opportunity. I will list some of the key action points here and refer to their book for an in-depth analysis (Gratton and Scott 2016).

- Pension system/long-term investment: people will have to work for longer and retirement age will need to increase. Work flexibility and a different balance of work/leisure time will need to be implemented to prevent burnout and exhaustion.
- A longer life cannot be structured in the traditional three stages (education/ work/retirement); a longer life will be multi-stage; education and employment will need to become more flexible; long-lived people will need to requalify over time.
- There will be a greater number of career transitions. Psychology and sociology have the task of understanding how people can make smoother career transitions.
- Architecture/infrastructure need to accommodate the growing proportion of older people: housing and city planning will need to accommodate the needs of older people.

Happily ever after?

Longevity is one of the greatest triumphs of humankind. The fact that we will all live longer raises many worries, however. Old age worries not only individuals, but also our collective psychology: people worry about what is going to happen to our planet, how our societies are going to function, given that the over-sixties are becoming the largest section of society.

However, not only is the pressure that the long-lived are perceived to present exaggerated, it can also be further reduced. There are of course other strategies by which being old could be made more desirable, but I have proposed that we should start with simple and realistic steps: dismantling some myths relating to longevity, understanding how diseases often associated with longevity may be effectively prevented (Bostrom 2005) and promoting intergenerational dialogue and solidarity.

This is not to say, of course, that individuals should bear the responsibility for their own health as they age. It is instead to say that policies that involve relatively minor costs could bring high benefits for all (older and younger). Healthy longevity is the result of a concerted effort that must happen at various levels, and stressing individual ability to control health is only one of them. Whether or not we, say, keep walking as we grow older depends not just on whether we have sufficient mobility, or whether we are willing to do so, but also on whether there is a set of infrastructures that allow us to feel safe in walking. The Brasilia Declaration on Ageing recommends that '[a]ll actions must . . . take into account the bio-physical, social, psychological, economic, and environmental determinants of health. Policies across sectors [of local government and its employees] must be coordinated and harmonized' (WHO 1996).

How we age, partly depends on luck of course; but it also depends on behavioural and environmental factors: luck aside, these can be controlled, but only through a coordinated effort (HM Government 2014: 5). Infra-structures and architectonic designs must reflect the findings relating to longevity. Safe areas for walking, well-lit streets, adequate pavements (together with pedestrian traffic lights that have sufficient duration for those with reduced mobility); support for community activities that encourage physical activity; provision of recreational services that offer elderly people exercise programmes that help them to maintain their mobility; the inclusion of information and education about longevity in the training programmes of health carers, social carers, recreational workers, city planners and architects; provision of pavements and cycle tracks close to residential areas to encourage walking or cycling as a part of daily activity; easy access to information about healthy longevity – these are just a few examples of cost-effective policies aimed at promoting *good longevity* (WHO 2015b).

Population ageing has not only raised mixed reaction and worries; it has also raised awareness of the need to protect older members of society from abuse and discrimination. One of the first documents on the rights of the older person was the Brasilia Declaration on Ageing (WHO 1996). One of the most recent was the Declaration of Rights for Older People in Wales (2014), and the United Nations is calling for a legally binding convention on the rights of older people (Office of the High Commissioner for Human

Rights 2014). Virtually all declarations and conventions on human rights contain statements about the fundamental human right of the elderly not to be discriminated against; for example, the Convention for the Protection of Human Rights and Fundamental Freedoms as amended by Protocol n. 11, 4 November 1950, art. 14, Prohibition of discrimination; or the European Social Charter (Revised) (3 May 1996), Part V, art. E.[3] The Charter of Fundamental Rights of the European Union (2000) prohibits:

> Any discrimination based on any ground such as sex, race, colour, ethnic or social origin, genetic features, language, religion or belief, political or any other opinion, membership of a national minority, property, birth, disability, *age* or sexual orientation. (art. 21, non-discrimination – my emphasis)

Universal ethical principles, however, remain abstract and empty concepts, mere unfulfilled ideals, unless policy directed at improving people's lives is implemented. When a policy involves relatively minor costs and high benefits for all (older and younger), it is clearly morally irresponsible not to endorse it.

The growing proportion of elderly people is a sign of the achievements of humankind in extending life and improving life conditions. This growth means that all of us get a better chance of living longer. This is to be celebrated as one of our greatest accomplishments, and the contribution that older people offer to society is to be appreciated, promoted and defended.

Notes

1　Further information on dementia can be found in other WHO documents that focus specifically on this topic at www.who.int/topics/dementia/en.
2　For example, in England the Department of Health has published a report (2011). See also HM Government (2014: 10). The American College of Sports Medicine (2010) also has various guidelines on healthy ageing, and recommends public policy to ensure promotion of physical activity as a public health measure. Wales has introduced, in 2014, the Active Travel Act, an Act of Parliament that requires local authorities to continuously improve facilities and routes for pedestrians and cyclists. This act requires local authorities to publish maps of safe walking and cycling routes and enhance these over time and new road schemes to consider the needs of pedestrians and cyclists (HM Government, 2014: 13).
3　Including also Universal Declaration of Human Rights 1948, Preamble; Council of Europe Convention for the Protection of Human Rights and Dignity of the Human Being with Regard to the Application of Biology and Medicine: Convention on Human Rights and Biomedicine 1997; United Nations Convention on Elimination of All Forms of Discrimination Against Women 1981.

References

Alecxih, L. Shen, L., Chan, I., Taylor, D., and Drabek, J. (2010), *Individuals Living in the Community with Chronic Conditions and Functional Limitations: A*

Closer Look, Washington DC: Office of the Assistant Secretary for Planning and Evaluation, Office of Disability, Aging and Long-Term Care Policy, US Department of Health and Human Services, www.aspe.hhs.gov/daltcp/reports/2010/closerlook.pdf (last accessed 17 October 2016).

American College of Sports Medicine (2010), 'Healthy aging: keeping active amid life changes', www.acsm.org/docs/fit-society-page/2010-fall-fspn_healthy-aging.pdf?sfvrsn=0 (last accessed 17 October 2016).

Bacher, B., Rice, R., Clasey, J., McCrory, J., and Harrison, A. L. (2002), 'The relationship between life-time physical activity and bone mineral density of the proximal femur', *Journal of Geriatric Physical Therapy*, 25.3: 38–9.

Backes-Gellner, U., and Veen, S. (2013), 'Positive effects of ageing and age diversity in innovative companies – large-scale empirical evidence on company productivity', *Human Resource Management*, 23.3: 279–95.

BBC (2016), 'A world without Down's syndrome', www.bbc.co.uk/iplayer/episode/b07ycbj5/a-world-without-downs-syndrome (last accessed 17 October 2016).

Bodnar, A., Ouellette, M., Frolkis, M., Holt, S. E., Chiu, C. P., Morin, G. B., Harley, C. B., Shay, J. W., Lichtsteiner, S., and Wright, W. E. (1998), 'Extension of life-span by introduction of telomerase into normal human cells', *Science*, 279.5349: 349–52.

Bonicelli, E., and Sciarretta, G. (2005), *La scienza e il sogno di vincere il tempo*, Milan: Raffaello Cortina.

Bostrom, N. (2005), 'The fable of the dragon tyrant', *Journal of Medical Ethics*, 31.5: 273–7.

Broccoli, L., Chong, L., Oelmann, S., Fernald, A. A., Marziliano, N., van Steensel, B., Kipling, D., Le Beau, M. M., and de Lange, T. (1997), 'Comparison of the human and mouse genes encoding the telomeric protein, TRF1: chromosomal localization, expression and conserved protein domains', *Human Molecular Genetics*, 6.1: 69–76.

Cheng, S., Suominen, H., Sakari-Rantala, R., Laukkanen, P., Avikainen, V., and Heikkinen, E. (1997), 'Calcaneal bone mineral density predicts fracture occurrence: a five-year follow-up study in elderly people', *Journal of Bone and Mineral Research*, 12.7: 1075–82.

Chief Medical Officer (2004), 'At least five a week: evidence on the impact of physical activity and its relationship to health', London, Department of Health.

Chubak, J., Ulrich, C. M., Tworoger, S. S., Sorensen, B., Yasui, Y., Irwin, M. L., Stanczyk, F. Z., Potter, J. D., and McTiernan, A. (2006), 'Effect of exercise on bone mineral density and lean mass in postmenopausal women', *Medicine & Science in Sports & Exercise*, 38.7: 1236–44.

Council of Europe Convention for the Protection of Human Rights and Dignity of the Human Being with regard to the Application of Biology and Medicine: Convention on Human Rights and Biomedicine 1997, Treaty No. 164, www.coe.int/en/web/conventions/full-list/-/conventions/treaty/164 (last accessed 18 October 2016).

Courtin, E., and Knapp, M. (2015), 'Social isolation, loneliness and health in old age: a scoping review', *Health and Social Care in the Community*, 25.3: 799–812.

Cruz-Jentoft, A. J., Baeyens, J. P., Bauer, J. M., Boirie, Y., Cederholm, T., Landi, F., Martin, F. C., Michel, J. P., Rolland, Y., Shneider, S. M., Topinkova, E., Vandewounde, M., and Zamboni, M. (2010), 'European Working Group on Sarcopenia in Older People. Sarcopenia: European consensus on definition and diagnosis: report of the European Working Group on Sarcopenia in Older People', *Age and Ageing*, 39.4: 412–23.

Dargent-Molina, P., Favier, F., Grandjean, H., Baudoin, C., Schott, A. M., Hausherr, E., Meunier, P. J., and Breart, G. (1996), 'Fall-related factors and risk of hip fracture: the EPIDOS prospective study', *The Lancet*, 348.9021: 145–9.

Declaration of Rights for Older People in Wales (2014), http://wales.gov.uk/docs/dhss/publications/140716olderen.pdf (last accessed 18 October 2016).

Department of Health, UK (2001), *National Service Framework for Older People*, London: Department of Health, www.gov.uk/government/uploads/system/uploads/attachment_data/file/198033/National_Service_Framework_for_Older_People.pdf (last accessed 17 October 2016).

Department of Health, UK, Chief Medical Officer (2004), At Least Five a Week: Evidence on the Impact of Physical Activity and its Relationship to Health: A Report from the Chief Medical Officer, London: Department of Health, www.ssehsactive.org.uk/sites/Exercise-Referral-Toolkit/downloads/resources/cmos-report-at-least-five-a-week.pdf (last accessed 17 October 2016).

Department of Health, UK, Chief Medical Officer (2011), *Start Active, Stay Active: A Report on Physical Activity for Health from the Four Home Countries*, London: Department of Health, www.gov.uk/government/uploads/system/uploads/attachment_data/file/216370/dh_128210.pdf (last accessed 17 October 2016).

Department for Transport, UK (2011), *Statistical Release, National Travel Survey*, 2011.

Department for Transport, UK (2012), *National Travel Survey: 2011*, London, www.gov.uk/government/uploads/system/uploads/attachment_data/file/35738/nts2011-01.pdf (last accessed 17 October 2016).

Dhital, A., Pey, T., and Stanford, M. R. (2010), 'Visual loss and falls: a review', *Eye*, 24.9: 1437–46.

Dong, X., Milholland, B., and Vijg, J. (2016), 'Evidence for a limit to human lifespan', *Nature*, doi:10.1038/nature19793 (last accessed 17 October 2016).

Epicurus (2017), *Letter to Menoeceus*, trans. Robert Drew Hicks, http://classics.mit.edu/Epicurus/menoec.html (last accessed September 2017).

EU Charter of Fundamental Rights (2000), www.europarl.europa.eu/charter/default_en.htm (last accessed 18 October 2016).

European Convention on Human Rights (1950), www.echr.coe.int/Documents/Convention_ENG.pdf (last accessed 18 October 2016).

European Social Charter (Revised) (1996), http://conventions.coe.int/Treaty/EN/Treaties/Html/163.htm (last accessed 18 October 2016).

Farley, A., McLafferty, E., and Hendry, C. (2011), *The Physiological Effects of Ageing*, Oxford: Wiley.

Franklin, B. A. (2008), 'Physical activity to combat chronic diseases and escalating health care costs: the unfilled prescription', *Current Sports Medicine Reports*, 7.3: 122–5.

Gianoudis. J., Bailey, C., Sanders, K., Nowson, C., Hill, K., Ebeling, P., and Daly, R. (2012), 'Osteo-cise: strong bones for life: protocol for a community-based randomised controlled trial of a multi-modal exercise and osteoporosis education program for older adults at risk of falls and fractures', *BMC Musculoskeletal Disorders*, 13.78, http://bmcmusculoskeletdisord.biomedcentral.com/articles/10.1186/1471-2474-13-78 (last accessed 17 October 2016).

Gillespie, L. D., Robertson, M. C., Gillespie, W. J., Sherrington, C., Gates. S., Clemson, L. M., and Lamb, S. E. (2012), 'Interventions for preventing falls in older people living in the community', *The Cochrane Database of Systematic Reviews*, www.bhfactive.org.uk/userfiles/Documents/Cochranereviewfalls.pdf (last accessed 17 October 2016).

GOS (2016), *Future of an Ageing Population*, www.ageing.ox.ac.uk/download/190 (last accessed 17 October 2016).

Gratton, L., and Scott, S. (2016), *The 100-Year Life*, London: Bloomsbury.

Hagberg, L. A., and Lindholm, L. (2006), 'Cost-effectiveness of healthcare-based interventions aimed at improving physical activity', *Scandinavian Journal of Public Health*, 34.6: 641–53.

Heidemeier, H., and Moser, K. (2009), 'Self–other agreement in job performance ratings: a meta-analytic test of a process model', *Journal of Applied Psychology*, 94.2: 353–70.

HM Government (2014), *Moving More, Living More: The Physical Activity Olympic and Paralympic Legacy for the Nation*, London: Cabinet Office, www.gov.uk/government/uploads/system/uploads/attachment_data/file/279657/moving_living_more_inspired_2012.pdf (last accessed 17 October 2016).

Hoops, M. L., Rosenblatt, N. J., Hurt, C. P., Crenshaw, J., and Grabiner, M. D. (2012), 'Does lower extremity osteoarthritis exacerbate risk factors for falls in older adults?', *Women's Health*, 8.6: 685–96.

Ikeda, H., Ishizaki, F., Shiokawa, M., Aoi, S., Iida, T., Chiho, C., Tamura, N., and Harada, T. (2012), 'Correlations between walking exercise and each of bone density, muscle volume, fluctuation of the center of gravity, and dementia in middle-aged and elderly women', *International Medical Journal*, 19.2: 154–7.

International Diabetes Federation (2013), *Managing Older People with Type 2 Diabetes: Global Guideline*, Brussels: International Diabetes Federation, www.ifa-fiv.org/wp-content/uploads/2014/02/IDF-Guideline-for-Older-People.pdf (last accessed 17 October 2016).

Kalisch, D. W., Aman, T., and Buchele, L. A. (1998), *Social and Health Policies in OECD Countries: A Survey of Current Programmes and Recent Developments*, Paris: Organisation for Economic Co-operation and Development.

Karlsson, M. K., Magnusson, H., von Schewelov, T., and Rosengren, B. E. (2013), 'Prevention of falls in the elderly – a review', *Osteoporosis International*, 24.3: 747–62.

Kingsley, D. E. (2015), 'Aging and health care costs: narrative versus reality', *Poverty Public Policy*, 7.1: 3–21.

Lee, W. K., Kong, K. A., and Park, H. (2012), 'Effect of preexisting musculoskeletal diseases on the 1-year incidence of fall-related injuries', *Journal of Preventive Medicine and Public Health*, 45.5: 283–90.

Longobardi, G. et al. (2012), 'Protective effect on mortality of physical exercise in adult and elderly patient with cardiovascular disease', *European Journal of Preventive Cardiology*, 19.1 SUPPL.1: S99.

Moreira, L. D., Oliveira, M. L., Lirani-Galvão, A. P., Marin-Mio, R. V., Santos, R. N., and Lazaretti-Castro, M. (2014), 'Physical exercise and osteoporosis: effects of different types of exercises on bone and physical function of postmenopausal women', *Arquivos Brasileiros de Endocrinologia & Metabologia*, 58.5: 514–22.

O'Connor, P. J., Aenchbacher III, L. E., and Dishman, R. K. (1993), 'Physical activity and depression in the elderly', *Journal of Aging and Physical Activity*, 1.1: 34–58.

OECD (2006), *Live Longer, Work Longer: A Synthesis Report*, Paris: Organisation of Economic Co-operation and Development.

Office of the High Commissioner for Human Rights (2014), 'UN Human Rights Chief offers her support for a new Convention on the rights of older persons', United Nations, www.ohchr.org/EN/NewsEvents/Pages/RightsOfOlderPersons.aspx (last accessed 18 October 2016).

Oliver, D., Foot, C., and Humphries, R. (2014), *Making our Health and Care Systems Fit for an Ageing Population*, London: King's Fund, www .kingsfund.org.uk/sites/files/kf/field/field_publication_file/making-health-care-systems-fit-ageing-population-oliver-foot-humphries-mar14.pdf (last accessed 17 October 2016).

Phillipson, C. (2015), 'The political economy of longevity: developing new forms of solidarity for later life', *Sociological Quarterly*, 56: 80–100.

Plato (1984), *Apology*, in *The Dialogues of Plato*, vol. I, *Euthyphro, Apology, Crito, Meno, Gorgias, Menexenus*, trans R. E. Allen, New Haven, CT: Yale University Press.

Pratt, M., Macera, C. A., and Wang, G. (2000), 'Higher direct medical costs associated with physical inactivity', *Physician and Sportsmedicine*, 28.10: 63–70.

Ramsey, P. (2009 [1970]), *Fabricated Man: The Ethics of Genetic Control*, New Haven, CT: Yale University Press.

Rantanen, T., Volpato, S., Ferrucci, L., Heikkinen, E., Fried, L. P., and Guralnik, J. M. (2003), 'Handgrip strength and cause-specific and total mortality in older disabled women: exploring the mechanism', *Journal of the American Geriatrics Society*, 51.5: 636–41.

Ribeiro, L. H. M., and Neri, A. L. (2012), 'Physical exercise, muscle strength and the day-to-day activities of elderly women', *Ciência & Saúde Coletiva*, 17.8: 2169–80, www.scielo.br/scielo.php?script=sci_arttext&pid=S1413-81232012000800027 (last accessed 17 October 2016).

Russo, A., Onder, G., Cesari, M., Zamboni, V., Barillaro, C., Capoluongo, E., Pahor, M., Bernabei, R., and Landi, F. (2006), 'Lifetime occupation and physical function: a prospective cohort study on persons aged 80 years and older living in a community', *Occupational and Environmental Medicine*, 63.7: 438–42.

Ruuskanen, J. M., and Ruopilla, I. (1995), 'Physical activity and psychological well-being among people aged 65 to 84 years', *Age and Ageing*, 24.4: 292–6.

Salzman, B. (2010), 'Gait and balance disorders in older adults', *American Family Physician*, 82.1: 61–8.

Seitz, D., Purandare, N., and Conn, D. (2010), 'Prevalence of psychiatric disorders among older adults in long-term care homes: a systematic review', *International Psychogeriatrics*, 22.7: 1025–39.

Shiroma, E. J., and Lee, I. M. (2010), 'Physical activity and cardiovascular health: lessons learned from epidemiological studies across age, gender, and race/ethnicity', *Circulation*, 122.7: 743–52.

Staudinger, U., and Bowen, C. (2011), 'A systemic approach to aging in the work context', *Zeitschrift für Arbeitsmarktforschung*, 44.4: 295–306.

Steinmo, S., Hagger-Johnson, G., and Shahab, L. (2014), 'Bidirectional association between mental health and physical activity in older adults: Whitehall II prospective cohort study', *Preventive Medicine*, 66: 74–9.

Studenski, S., Perera, S., and Patel, K. (2011), 'Gait speed and survival in older adults', *Journal of the American Medical Association*, 305.1: 50–8.

Sugiswawa, S., Liang, J., and Liu, X. (1994), 'Social networks, social support and mortality among older people in Japan', *Journals of Gerontology*, 49.1: S3–13.

Suominen, H. (2006), 'Muscle training for bone strength', *Aging Clinical and Experimental Research*, 18.2: 85–93.

Terraneo, M. (2015), 'Inequities in health care utilization by people aged 50+: evidence from 12 European countries', *Social Science and Medicine*, 126: 154–63.

United Nations Convention on Elimination of All Forms of Discrimination against Women (1981), www.un.org/womenwatch/daw/cedaw (last accessed 18 October 2016).

Universal Declaration of Human Rights (1948), www.un.org/Overview/rights.html (last accessed 18 October 2016).

US Department of Health and Human Services (2008), *Physical Activity Guidelines Advisory Committee Report*, Washington DC: Department of Health and Human Services.

Wang, F., McDonald, T., Champagne, L. J., and Edington, D. W. (2004), 'Relationship of body mass index and physical activity to health care costs among employees', *Journal of Occupational and Environmental Medicine*, 46.5: 428–36.

Weisstub, D. N. (2015), 'Longevity ethics', *Medicine and Public Health*, 1.1: 148–50.

Weuve, J., Kang, J. H., Manson, J. E., Breteler, M. M., Ware, J. H., and Grodstein, F. (2004), 'Physical activity, including walking and cognitive function in older women', *Journal of the American Medical Association*, 292.12: 1454–61.

WHO (1996), *The Brasilia Declaration on Ageing*, Geneva: World Health Organization, www.oneworld.org/helpage/info/brasilia.html (last accessed 17 October 2016).

WHO (1998a), *Life in the 21st Century: A Vision for All*, Geneva: World Health Organization, www.who.int/whr/1998/en/whr98_en.pdf (last accessed 17 October 2016).

WHO (1998b), *Growing Older, Staying Well: Ageing and Physical Activity in Everyday Life*, Geneva: World Health Organization, http://apps.who.int/iris/bitstream/10665/65230/1/WHO_HPR_AHE_98.1.pdf (last accessed 17 October 2016).

WHO (2000), *Home-based Long-term Care: Report of a WHO Study Group*, WHO Technical Report Series 898, Geneva: World Health Organization, http://whqlibdoc.who.int/trs/WHO_TRS_898.pdf (last accessed 17 October 2016).

WHO (2002a), *Active Ageing: A Policy Framework*, Geneva: World Health Organization, http://apps.who.int/iris/bitstream/10665/67215/1/WHO_NMH_NPH_02.8.pdf (last accessed 17 October 2016).

WHO (2002b), 'Healthy aging is vital for development', press release, WHO/24, www.who.int/mediacentre/news/releases/release24/en (last accessed 17 October 2016).

WHO (2010), *Global Status Report on Noncommunicable Diseases*, Geneva: World Health Organization, www.who.int/nmh/publications/ncd_report_full_en.pdf (last accessed 17 October 2016).

WHO (2011), *Global Health and Aging*, NIH Publication no. 11–7737, Geneva: World Health Organization, www.who.int/ageing/publications/global_health.pdf (last accessed 17 October 2016).

WHO (2012), *Good Health Adds Life to Years: Global Brief for World Health Day 2012*, Geneva: World Health Organization, http://apps.who.int/iris/bitstream/10665/70853/1/WHO_DCO_WHD_2012.2_eng.pdf (last accessed 17 October 2016).

WHO (2014), 'Aging and life course: facts about aging', www.who.int/ageing/about/facts/en (last accessed 17 October 2016).

WHO (2015a), Global Strategy and Action Plan on Ageing and Health (2016–2020): A Framework for Coordinated Global Action by the World Health Organization, Member States, and Partners across the Sustainable Development Goals, Geneva: World Health Organization, www.who.int/ageing/GSAP-Summary-EN.pdf?ua=1 (last accessed 17 October 2016).

WHO (2015b), *World Report on Aging and Health*, Geneva: World Health Organization, http://apps.who.int/iris/bitstream/10665/186463/1/9789240694811_eng.pdf?ua=1 (last accessed 17 October 2016).

WHO (2017), 'Noncommunicable diseases fact sheet', www.who.int/mediacentre/factsheets/fs355/en (last accessed 2 May 2017).

WHO (n.d.), 'Health topics: dementia', www.who.int/topics/dementia/en (last accessed 17 October 2016).

3

Reproductive technologies and the family in the twenty-first century

Daniela Cutas and Anna Smajdor

The first IVF baby was born in 1978 in the UK, following an intervention that had not been preceded by any clinical trials. After Louise Brown's birth, legislators and policymakers rushed to create an ethico-legal framework within which this new development could be practised without outraging public sensibilities. Since then, the speed and direction of scientific research as well as the practice and regulation of reproductive technology have been inexorably shaped by assumptions concerning family, fertility and reproduction. Research towards ever more sophisticated medical technologies for the purpose of the relief of infertility has raised relatively few concerns, provided the procedures were proven to be satisfactorily safe, and insofar as they were used to facilitate and reinforce existing norms about family structure and relationships. Ideas of what a family is (or should be) have a powerful influence on determining which potential technological innovations in human reproduction are developed and funded, and who can access them.

Social, legal and biological parenthood did not invariably coincide in the past. Different jurisdictions have various approaches to the ascription of parental rights and responsibilities. However, the default legal position is that a woman who gives birth to a child is that child's mother and her husband is the father – regardless of whether she is the genetic mother or he the genetic father. Embedded in this view is the expectation that the two members of the married couple are the legal and social parents and also the biological parents of the child. These legal measures have not been extended in the same way to the case of same-sex married couples. Furthermore, the child's 'right to know' her genetic origins, which features as a core argument for transitioning from anonymous to non-anonymous gamete donation in many countries, has not persuaded legislators that children who are born to married couples via sexual reproduction should also be aware of who their genetic father is (Ravelingien and Pennings 2013). These assumptions indicate that biological relationships are subservient to the nuclear family in the eyes of the law – and furthermore, specifically to the heterosexual nuclear family (Cutas and Chan 2012).

Developments in reproductive technologies have created new types of connections. A woman can now give birth to a child who is not genetically *her own*, or to a child who bears genetic material from both herself and another woman (via, for example, mitochondrial DNA transfer). Perhaps, in the future, a woman will be able to give birth to a child who is *only her own genetically* (via solo reproduction with *in vitro*-created gametes or human cloning). Gamete donation and surrogacy bring about challenges that many legislatures are still trying to sort out. Changes brought about by such innovations are increasingly hard to reconcile with still pervasive nuclear family expectations in ethics and regulation. If not so long ago, legal parenting at least approximately mirrored genetic parentage (one female mother and one male father, preferably married to each other), the diversification of genetic parentage itself, alongside a host of other sociocultural changes, pushes the model further into what some have called the *crisis* of the family (Collier 1999: 38–58; Cutas and Chan 2012; Szczurek 2013: 355–90).

In this chapter, we explore the nature and significance of these interactions in the context of research towards prospective reproductive technologies. We will start by giving a brief overview of the aftermath of the advent of IVF in the UK. We will then discuss how desires come to be classified as medical needs, and the relation between reproductive aspirations, the circumstances of those who hold them, and the weight that they are given. This will lead us to a discussion of current definitions of infertility, which have constituted the core criteria of access to reproductive technologies. These definitions determine not only who has access to reproductive technologies, but also whose desires are deemed worthy to motivate research efforts to make these treatments possible.

The regulatory background in the UK

After Louise Brown's birth was reported in the British press, amid much speculation and anxiety about the likely effects of IVF, the government of the time concluded that some regulatory intervention would be required. The public needed to know that scientists' and doctors' activities were subject to ethical and legal oversight (Deech and Smajdor 2007). A report into the ethics of IVF and embryo research was commissioned: its task was to answer the question of whether the use of IVF was acceptable and if so, in which circumstances – and to advise on the acceptability of using human embryos in research. Mary Warnock, the chair of the committee, held the position that legal and regulatory barriers need not be based solely on the avoidance of adverse consequences. Rather, she claimed that these barriers reflected a public desire for boundaries, and served to express and enforce moral beliefs that define our social identity:

> [i]n recognising that there should be limits, people are bearing witness to the existence of a moral ideal of society. (Report of the Committee of Inquiry into Human Fertilisation and Embryology 1984: 2)
> A society which had no inhibiting limits, especially in the areas with which we have been concerned . . . would be a society without moral scruples. *And this nobody wants.* (Report of the Committee of Inquiry into Human Fertilisation and Embryology 1984: 2) [our emphasis]
> Barriers, it is generally agreed, must be set up. (Report of the Committee of Inquiry into Human Fertilisation and Embryology 1984: 3)

While later reports were critical of Warnock's approach, and less enthusiastic about the importance of barriers (House of Commons Science and Technology Committee 2004–05: 4–5), it remains the case that the direction of scientific progress, and the form of the post-reprotec family, have been profoundly shaped by these regulatory constraints. Since the UK's legislation in this context formed a blueprint for many other countries (Rasmussen 2004; Bennett 2008), its assumptions and the barriers based on them have had an impact worldwide on what can and cannot be done with reproductive technology.

The Warnock Report was published in 1984. Marriage and couples feature extensively in its pages. Although considering arguments in favour of individual women and men who might wish to become parents with the help of reproductive technologies, the Committee members held the belief 'that as a general rule it is better for children to be born into a two-parent family, with both father and mother' (Report of the Committee of Inquiry into Human Fertilisation and Embryology 1984: 11–12). The legislation based on the Warnock Report's findings, the Human Fertilisation and Embryology Act, was not passed until 1990 (subsequently amended in 2008). Globally, IVF and related techniques had grown and proliferated during this period, opening many new possibilities that had not been envisaged previously. In view of this, it is not all that surprising that some think the Act was already outdated almost from the moment it became law (Lee and Morgan 2001: 2). Many new and prospective avenues for scientific and medical exploration, such as surrogacy, pre-implantation diagnosis, mitochondrial donation, egg freezing and the possibility of creating gametes *in vitro*, were emerging. The latter is of particular interest because prospective innovations such as *in vitro*-created gametes or human cloning may disintegrate the biological boundaries that currently help to define what we mean by reproduction (Hendriks et al. 2015).

However, the regulatory framework is only part of the story here. Another issue is medicine itself. From its very earliest days the relationship between reproductive technology and medicine has been uneasy and fraught with controversy. The legislative framework gives a broad overview of what can and cannot be undertaken in a general sense. But the medical establishment determines which patients are eligible for treatment – and which needs require medical intervention or research into the development of future treatment options. As we will show, the clinical basis for these is very far removed from many other areas of medicine. Although the use of reproductive technologies is thought of and presented as 'treatment' for infertility, very often the patient may be in perfect reproductive health, but have a partner who has fertility problems – or simply not have succeeded to become a parent in the natural way with a specific partner. In reproductive medicine, one person can be the vehicle by which another person's medical problem is addressed. Moreover, the aim of treatment is not to remedy the fertility problem (through repairing fallopian tubes, for example) but to provide people with a baby. This means it is not always easy to see what makes one procedure a legitimate medical treatment, and not others.

IVF is accepted in many countries across the world as a legitimate – and on the whole a morally acceptable – means of treating infertility. Reproductive cloning, on the other hand, is stringently forbidden almost everywhere in the world, and is regarded by many as being morally abhorrent (Wellcome Trust 1998). Yet both IVF and reproductive cloning are techniques which could enable a person to have genetically related offspring. Interestingly, surveys of public views about the permissibility of reproductive cloning indicate that cloning is felt to be more acceptable if it is presented as an option for a heterosexual couple (as opposed to e.g. a single individual) (Shepherd et al. 2007). That is, it is the perception of a technology as a means of reproducing the nuclear family which seems to facilitate its acceptance – while at the same time what counts as a medical problem depends upon how close one's circumstances are to the preferred relationship form.

Normative assumptions about family structure are also mirrored in medical practice and in regulatory approaches to fertility treatment. In many countries, access to treatment depends heavily on the degree to which the patient matches a specific relationship and/or family model. Fertility treatment is more readily accepted when it facilitates the creation of nuclear families. This is still how things are in many jurisdictions: though more and more exceptions are made. For example, in the Netherlands, single women are often refused treatment. Those who do receive treatment are required to undergo psychological tests (Pieters 2015). In Sweden, female same-sex couples only acquired the right to access fertility treatments in 2005, and single women in 2016. As we will see further on in this chapter, in Italy (Legge 40, 2004) and France (Loi n° 2011–814) only heterosexual couples are allowed access. This means that although different sets of individuals may have a similar desire – to have a child – that desire is more likely to be construed as a legitimate medical *need* when the individual is in a partnership with someone from the other sex. Similarly, when we look at access to uterus transplantation, we can see that while many people may have the desire to gestate a baby in their own body, only those who fit certain norms can access treatment that would enable them to do so.

Reproductive needs and the nuclear family

Reproductive technologies tend to be allowed to operate only within boundaries that are perceived to replicate or repair the natural. Sheila Jasanoff notes that

> genetic engineering threatens or calls into question many of the categories that have been accepted as foundational in the ordering of societies, both ancient and modern. These include the fundamental divisions between nature and culture, moral and immoral, safe and risky, god-given and human-made. (2005: 26)

The ways in which reproductive technologies are regulated illustrate some attempts to limit their 'disruptive potential', to keep them within conceptual boundaries that match certain norms about what constitutes the family. In what follows we will show how, if we look for the reinforcement of concepts of nature and biology in the family, we can see how deeply entrenched these ideas are, and how powerful a role they play in managing the potential threats inherent in new technologies. Nuclear families in this context are not necessarily favoured for overtly political or even moral reasons, but because they represent a natural or biological norm. As Barrett and McIntosh put it:

> It is in the realm of gender, sexuality, marriage and the family that we are collectively most seduced by appeals to the natural. In this realm the shifting norms of practice are solidified, some to be sanctified and others condemned. The prevailing form of family is seen as inevitable, as naturally given and biologically determined. (1991: 27)

This 'naturalness' of the nuclear family form has been contested by researchers from various angles (Stacey 2011; Diduck and Kaganas 2012). It is clear that when the nuclear family is used as the measure of how things should be, this will have an impact on what research is encouraged and seen as responding to a legitimate need. This has already been seen in the case of IVF. It seemed evident to the Warnock Committee that IVF's primary use should be to facilitate the formation of nuclear families. They foresaw that single women or perhaps lesbian couples might want to use IVF with sperm donation and in doing so, to create children who would be raised without a father. In an effort to circumvent this, a clause was included in the law governing fertility treatment, requiring clinics to 'consider the need of a child for a father'. Effectively, this gave clinics grounds for refusing single women and lesbian couples, since fertility doctors had a legal obligation to consider the welfare of the future child, and since this welfare consideration was explicitly linked with the need for a father. In this way, the primacy of the nuclear family was protected, at least to some degree.

It is clear that genes play a very important part in our understandings of reproduction. However, although genetic connections are often viewed as being determinative of parenthood, as we will see further in this chapter when we discuss mitochondrial DNA transfer, it is now becoming possible to ask whether a genetic link is either necessary or sufficient to justify calling a child one's own 'biological' offspring (Barritt et al. 2001). Following from this, we can ask whether any or all technologies which can produce children genetically related to some specific individual should be categorised as fertility treatments. For example, the World Health Organization (WHO) (1998) described reproductive cloning as replication rather than reproduction and stated that because it was not reproduction it should therefore be prohibited. Human reproduction, it has been argued, is essentially collaborative and sexual, while cloning is not sexual and can be non-collaborative, and thus is more akin to manufacture than reproduction (National Bioethics Advisory Commission 1997). The relationship between genes, parenthood and reproduction also feeds into the categorisation of certain technological possibilities as being therapeutic, curative, or needs related. Many people want babies related to them in particular ways, but only some of these people are deemed to have needs that can or should be addressed through medical technology. For example, same-sex male couples and single men cannot access treatment in the UK or in Sweden, because of the illegality of surrogacy. A number of countries impose strict cut-off ages for women wishing to access treatment. Denmark, Belgium and the Netherlands permit IVF only for women who are under 45, while France requires that women must be 'of reproductive age' (Brigham et al. 2013). While the desire for a baby may be the same in all cases, only some people's desires are regarded as being suitable for fulfilment through medical treatment. The categorisation of desires as needs often follows a particular kind of a biological model, so that although we use technology to supersede and transcend biology in various ways, other possibilities are marked off as taboo.

An interesting example of the relationship between genes, family norms and research/treatment priorities is the case of mitochondrial DNA transfer. Those who seek this procedure may be at risk of transmitting a mitochondrial mutation to their children. This risk is removed when the mitochondrial DNA of the carrier prospective mother is replaced with another woman's. Using this technique, the resulting offspring inherits genetic material from three people: the woman who provides the nucleus of the egg, the woman who provides the mitochondrial DNA and the man who provides the sperm. The presence of DNA from three different people has made the technique controversial. It has often been referred to in the press as creating babies with three parents (Hamzelou 2016; Macrae 2016).

However, many scientists in the field insist the term 'three-parent baby' is inaccurate because the nuclear DNA used in the child's creation is still

from two people only (Sample 2015). The manoeuvre to exclude the mitochondrial DNA contributor from the status of genetic parent makes it easier
to construe mitochondrial DNA transfer as therapeutic. Rather than
intervening to add genes or progenitors to a process that naturally involves
only two genetic contributors, the addition of mitochondrial DNA is classified
as 'mending' or 'replacing' a faulty part in the maternal egg. Thus, according
to this approach, the child still has only one genetic mother and the process
is acceptable as it is reconfigured to maintain our expectations of the biological norm: each child has exactly two genetic parents.

This reframing to angle the procedure away from anything as drastic as
introducing a third genetic parent into reproduction has given the technique
a kind of validation, and this enables the related research on which it is
based to progress. This is achieved, however, by dint of redefining what
we mean by genetic parenthood – a not insignificant development. The
exclusion of one of the DNA contributors from parent status seems to open
the possibility that genetic parenthood is a matter of *degree*, since the
mitochondrial donor simply does not provide *enough* genetic material in
relation to the other parents. While we do not seek to define genetic parenthood here, it is clear that the question of what constitutes a parent is one
that is *raised*, rather than resolved, by the development of such techniques.
Yet their permissibility seems to depend on the redefinition of parenthood
so as to forestall this question.

Mitochondrial DNA transfer also opens up other possibilities for those
who wish to have children in non-traditional families. For example, mitochondrial DNA transfer could be used to 'rejuvenate' the eggs of women
who may already be past the optimal reproductive age. It has been suggested
that replacing the mitochondrial DNA of the older woman's egg with that
of a younger woman can 'improve' the quality of the egg, thus increasing
the chances of a successful pregnancy (Reynier et al. 2001; May-Panloup
et al. 2005; Wolf et al. 2015). However, the use of mitochondrial transfer
for this purpose has to date not been licensed by the HFEA in the UK.

We have focused almost entirely on aspects of genetic parenthood. However,
there are of course other ways in which people can become biological
parents. One of these is gestation. Women who seek medical treatment in
order to become mothers may value different aspects of motherhood differently. One might wish to transmit genes to her child, another might wish
to mature the gametes in her own body following ovarian tissue transplant,
while others might place a higher value on gestating a child. Again, these
possibilities reveal differences in the way that the desires of some individuals
are integrated into the discourse of medical need and treatment, while those
of others are excluded. We will look here at uterus transplants as a particularly
interesting example.

Children have recently been born after having been carried in a uterus
transplanted into a woman from another woman (*The Guardian* 2015).
The paradigmatic patient for this treatment is one who was born without

a uterus. However, only a very specific subset of such individuals are selected to receive these transplants. Such transplants have been offered as part of a clinical trial. In this scenario, the boundary between therapy and research is blurred. For the women involved, their only chance of access to this treatment may be enrolment on the trial. Because regulations for research are tighter than those for standard treatment options, those undertaking the trials are able to impose narrower constraints on eligibility than might otherwise be the case.

According to one group of experts, the recipient must be 'a genetic female of reproductive age'. She must have 'a personal or legal contraindication to surrogacy and adoption'. Her wish to undergo uterine transplant must not be 'irrational'. She must not 'exhibit frank unsuitability for motherhood' (Lefkowitz et al. 2012). Another research group stipulate that to be eligible, patients must be:

- Women with AUI [absolute uterine infertility]
- Ages 20–35, with working ovaries
- BMI of less than 30
- Cancer free for at least 5 years
- Negative for HIV, hepatitis B and C, chlamydia, gonorrhoea and herpes
- No history of diabetes
- Non-smoker (Baylor Scott and White Health 2017).

It is interesting to note the variations in these criteria. A potential recipient has much more to prove than simply having been born without a uterus. Some of these requirements in fact seem to have little to do with the biological condition of lacking a uterus, at all. This, it has been argued, is unfair (Murphy 2015): if a woman should be eligible for uterus transplant because she was born without a uterus, then transwomen, all of whom were born without a uterus, should be an obvious category to be included here. Moreover, some men might genuinely wish to experience pregnancy: why would women's longing to experience pregnancy be worthy of considerable expense and not men's?

One obvious answer might be risk. It is easy to assume that to circumvent the natural biological boundaries of reproduction will be extraordinarily and excessively risky. However, when one examines the question more closely it becomes evident that risk cannot be the whole answer here. For one thing, uterus transplantation in itself is *extremely* risky. When live donation is undertaken, significant medical harms are imposed on two previously healthy women. Although transplantation to a male body would indeed be complex, it is not obvious that it would be so much riskier than female-to-female transplantation that it cannot be countenanced. One of the teams involved in the recent transplant trials commented that they had foreseen possible interest from the trans community – and had deliberately drawn up their protocol so as to exclude its members despite the fact that they admitted such transplants would be feasible (Grady 2015).

Expectations of what is 'natural' for a woman and a man (and what is or is not a 'genetic female'), respectively, determine whose desires will be appreciated and will motivate research efforts. Furthermore, one could claim that the fact that the pain and suffering caused by pregnancy only befall women is a natural inequality that we should aim to correct (Smajdor 2007; 2012): this would provide yet another motivation for also pursuing research aimed at providing uterus transplants for men.

An alternative angle might be that reproductive technologies should be used only to recreate specific sorts of relationships between the child and the parent, which IVF fulfils (within the nuclear family parameters), but reproductive cloning or womb transplants for men do not. An associated concern may be that some interventions do not really meet a medical need, while others do: for example, if a treatment is being offered to someone who is biologically infertile, as opposed to someone who is not (*The Lancet* 2001). This distinction is tricky, however, as we will see in our next section, because infertility itself is defined in a particular way that may exclude medical conditions in single individuals but include unexplained non-conception in heterosexual couples. We will show how applying such criteria could help to pick out some possibilities while excluding others.

Defining infertility

The WHO defines infertility as a disease of a sexually active couple. Thus, infertility is 'a disease of the reproductive system defined by failure to achieve a clinical pregnancy after 12 months or more of regular unprotected sexual intercourse' (Zegers-Hochschild et al. 2009). What is implicit in this definition is that the unprotected sexual intercourse here is heterosexual. Clearly, a same-sex couple will not achieve a clinical pregnancy, however many months they have unprotected sexual intercourse. The heterosexual couple *should* in the natural course of events be able to achieve a pregnancy in this way. This way of defining infertility may not be consistent with intuitions about what infertility is. For example, if Jane lacks ovaries, we might think of her as being infertile. However, she is not infertile according to the WHO's current definition – unless she has had unprotected sexual intercourse with a man over the duration of twelve months.

The WHO's definition is influential in determining how infertility is defined in specific countries. For example, in the UK, the National Institute for Health and Care Excellence includes the following recommendation: 'A woman of reproductive age who has not conceived after 1 year of unprotected vaginal sexual intercourse, in the absence of any known cause of infertility, should be offered further clinical assessment and investigation along with her partner' (2013). Because a requisite period of heterosexual intercourse is often a criterion of access to fertility treatment as well as to the medical investigations that precede any intervention, unless she fulfils this condition, she might not even find out that she lacks ovaries. Accordingly, the eligibility of patients (i.e. their *need* for treatment) has come to be dictated in part by a social factor rather than a clinical one: whether or not the patient is infertile depends on whether she has a particular sort of relationship. Provided that claimants are heterosexual, a construction of need can be mapped onto the underlying biological norm.

Clearly, though, it is possible that a lesbian or single woman could suffer from blocked fallopian tubes, or any other reproductive pathology. However, the reproductive pathology is implicitly of lesser importance than the structure of the relationship within which the patient finds herself. In practice, while in some countries definitions such as the WHO one cited above have been amended to allow same-sex couples or single individuals access to fertility treatments, in others they have been made more restrictive to include e.g. the provision that the members of the couple have to be married or cohabit, and of a certain age (not too old).

According to French law:

> the object of reproductive medical assistance is to remedy the infertility of a couple or to avoid the transmission to the child or a member of the couple, of a disease of particular gravity. The pathological character of infertility has to be medically diagnosed. The man and the woman have to be alive, of reproductive age, and had previously consented to the embryo transfer or the insemination. (Loi n° 2011–814)

Likewise, Italian law restricts eligible infertility to infertility in heterosexual couples. The purpose of assisted reproductive technologies (ARTs) is to 'help solve the reproductive problems caused by human sterility or infertility . . . when there are no other therapeutic means to remove the causes of sterility or infertility' (Legge 40, 2004, art. 1). However, the law specifically stipulates that access to such technologies is limited to 'adult couples of different sex, married or cohabiting, of reproductive age, both living' (Legge 40, 2004, art. 5).

A draft bill recently proposed in India would see access to surrogacy restricted to heterosexual couples married for at least five years (Srivastaval 2016). This preference for the (preferably married or at least cohabiting) heterosexual couple, in addition to its *natural*, assumed procreative potential, has been viewed as essential for the interests of the child: as we have seen above, this interest has been invoked in UK regulations of access to fertility treatment. It is assumed to be in children's interests to have a mother and a father. A growing body of empirical research indicating that what matters most for children's well-being seems to be the quality of family relationships rather than family form (Golombok 2015) is at odds with these expectations. Specifically, the 'no difference' outcome for children from being raised by same-sex as opposed to different-sex parents has been deemed to have reached consensus in the literature (Adams and Light 2015). Hence, if it is the interests of children that is the paramount concern here, directing research and treatment preferably or uniquely towards fulfilment of heterosexual partners' desires is questionable. The UK's requirement in terms of a child's need for a father was removed in 2008 – following decades of controversy. The belief of the members of the Warnock Report, that being raised by a mother and a father is better for children, is understandable in the context of its time and pre-dates relevant findings from empirical research: however, three decades on, the same belief has become much harder to support.

The disease of infertility, as defined above, is contextual to a degree far greater than most other diseases. Fertility treatments do not aim at restoring the capacity to achieve a pregnancy via sexual intercourse. Indeed, having become a parent via fertility treatments, one might not have reproduced at all (if gamete donors were used), and at the end of the treatment a couple might be just as incapable of achieving a clinical pregnancy as before. Simone Bateman (2002) argues that medicine should only seek to correct medical pathologies. Another perspective is taken by Habbema et al. (2004) who suggest that it is the symptom of non-conception that is addressed by fertility treatment. However, although non-conception *can* be a symptom which is treated in fertility medicine, it is not always necessarily the main symptom which brings people to the clinic. For example, those who want to use pre-implantation genetic diagnosis (PGD) or mitochondrial transfer to avoid having a child with a particular gene or a mitochondrial condition, may be very well able to reproduce unaided. It is just that they cannot have the *sort* of child they want, with the *person* they want to have it with, without medical help.

If we consider whether there *is* a specific necessary symptom associated with all claimants for fertility treatment, it becomes evident that there is a shared feature: *the aspiration to have a (specific kind of) child* (Smajdor and Cutas 2015). A woman may have blocked fallopian tubes for years but not become aware of this until she decides to reproduce. The suffering of infertility is dependent on having a specific desire. It is the desire that is treated, regardless of whether the patient is suffering from a reproductive pathology. Without this desire, infertility does not require treatment: on the contrary, it is convenient. Since it is the reproductive aspiration that is treated, and if family form is not determinative of children's well-being, then there seems to be no reason why heterosexual couples' reproductive aspirations should count as medical needs any more than those of other categories of individuals.

Conclusion

Embedded and unquestioned assumptions about reproduction, fertility and the family determine whose claims and whose suffering deserve attention. Widely shared beliefs, in their turn, contribute to this effect, and legal and professional regulations solidify it. As access to research and clinical trials can be influenced by these unquestioned assumptions, we can see the ongoing cycle of normative assumptions that feed into notions of need, which then generate treatments, for which eligibility is constrained by norms, which are then contested – partly on the basis of the new possibilities that have emerged.

The relationships between reproduction, family, medical need and research, are such as to grate against one another. Each new possibility poses new challenges to the ways in which we conceptualise the family. One possible response to this is to attempt to enforce and solidify normative expectations about the family and appropriate roles within it, and to protect it from the threats of innovation. Another might be to ask whether the nuclear family really merits such concern and protection – in a way in which other family types do not. We hope in this chapter to have highlighted some of the tensions between, on the one hand, science's potential to undermine assumptions and expectations in the realm of family-making, and on the other, the potential of these assumptions and expectations to influence the direction of research.

References

Adams, J., and Light, R. (2015), 'Scientific consensus, the law, and same sex parenting outcomes', *Social Science Research*, 53: 300–10.

Barrett, M., and McIntosh, M. (1991), *The Anti-Social Family*, London: Verso.

Barritt, J., Brenner, C. A., Malter, H. E. et al. (2001), 'Mitochondria in human offspring derived from ooplasmic transplantation', *Human Reproduction*, 16.3: 513–16.

Bateman, S. (2002), 'When reproductive freedom encounters medical responsibility: changing conceptions of reproductive choice', in E. Vayena, P. J. Rowe and P. D. Griffin (eds), *Current Practices and Controversies in Assisted Reproduction: Report of a WHO Meeting*, Geneva: World Health Organization, 320–32.

Baylor Scott and White Health (2017), 'Womb transplantation', baylorhealth.com/AdvancingMedicine/AreasOfResearch/Transplantation/Pages/WombTransplantation.aspx (last accessed February 2017).

Bennett, B. (2008), *Health Law's Kaleidoscope: Health Law Rights in a Global Age*, Aldershot: Ashgate.

Brigham, K. B., Cadier, B., and Chevreul, K. et al. (2013), 'The diversity of regulation and public financing of IVF in Europe and its impact on utilization', *Human Reproduction*, 1.28.3: 666–75.

Collier, R. (1999), 'Men, heterosexuality and the changing family: (re)constructing fatherhood in law and social policy', in G. Jagger and C. Wright (eds), *Changing Family Values*, London: Routledge, 38–58.

Cutas, D., and Chan, S. (eds) (2012), *Families – Beyond the Nuclear Ideal*, London: Bloomsbury.

Deech, R., and Smajdor, A. (2007), *From IVF to Immortality*, Oxford: Oxford University Press.

Diduck, A., and Kaganas, F. (2012), *Family Law, Gender and the State. Text, Cases and Materials*, Oxford: Hart.

Golombok, S. (2015), *Modern Families. Parents and Children in New Family Forms*, Cambridge: Cambridge University Press.

Grady, D. (2015), 'Will uterine transplants make male pregnancy possible?', *New York Times*, 16 November, www.nytimes.com/2015/11/16/insider/will-uterine-transplants-make-male-pregnancy-possible.html (last accessed May 2017).

Guardian (2015), 'Baby born from grandmother's donated womb', 25 August, theguardian.com/society/2015/aug/25/baby-born-from-grandmothers-donated-womb (last accessed March 2017).

Habbema, J. D., Collins, J., and Leridon, H. et al. (2004), 'Towards less confusing terminology in reproductive medicine: a proposal', *Fertility and Sterility*, 82.1: 36–40.

Hamzelou, J. (2016), 'Exclusive: World's first baby born with new "3 parent" technique', *New Scientist*, 27 September, www.newscientist.com/article/210721 9-exclusive-worlds-first-baby-born-with-new-3-parent-technique (last accessed May 2017).

Hendriks, S., Dancet, E. A., van Pelt, A. M. et al. (2015), 'Artificial gametes: a systematic review of biological progress towards clinical application', *Human Reproduction Update*, 21: 285–96.

House of Commons Science and Technology Committee (2004–05), *Human Reproductive Technologies and the Law. Fifth Report of Session 2004–05*, vol. I, HC 7–1, London: Stationery Office.

Jasanoff, S. (2005), *Designs on Nature*, Princeton, NJ: Princeton University Press.

Lancet (2001), Editorial, 'Biological uncertainties about reproductive cloning', 358.9281: 519.

Lee, R. G., and Morgan, D. (2001), *Human Fertilisation and Embryology*, London: Blackstone.

Lefkowitz, A., Edwards, M., and Balayla, J. (2012), 'The Montreal criteria for the ethical feasibility of uterine transplantation', *Transplant International*, 25.4: 439–47.

Legge 40 (2004), 19 February, camera.it/parlam/leggi/04040l.htm (last accessed March 2017).

Loi n° 2011–814 du 7 juillet 2011 relative à la bioéthique, art. 33, legifrance.gouv. fr (last accessed March 2017).

Macrae, F. (2016), 'Britain's first three-parent baby could be born within one year: scientists say the controversial IVF technique is ready for use', *Daily Mail*, 9 June, www.dailymail.co.uk/sciencetech/article-3631770/Britain-s-three-parent-baby-born-ONE-YEAR-Scientists-say-controversial-IVF-technique-ready-use.html (last accessed May 2017).

May-Panloup, P., Chretien, M. F., and Jacques, C. et al. (2005), 'Low oocyte mitochondrial DNA content in ovarian insufficiency', *Human Reproduction*, 20.3: 593–7.

Murphy, T. (2015), 'Assisted gestation and transgender women', *Bioethics*, 29.6: 389–97.

National Bioethics Advisory Commission (1997), *Cloning Human Beings: Report and Recommendations*, bioethicsarchive.georgetown.edu/nbac/pubs/cloning1/cloning.pdf (last accessed February 2017).

National Institute for Health and Care Excellence (2013), *Clinical Guideline [CG156]. Fertility Problems: Assessment and Treatment* (updated 2016), nice.org.uk/guidance/cg156/chapter/Recommendations#defining-infertility (last accessed March 2017).

Pieters, J. (2015), 'Single women routinely refused IVF treatment', NLTimes.nl, 22 July, nltimes.nl/2015/07/22/single-women-routinely-refused-ivf-treatment (last accessed March 2017).

Rasmussen, C. (2004), 'Canada's Assisted Human Reproduction Act: is it scientific censorship, or a reasoned approach to the regulation of rapidly emerging reproductive technologies', *Saskatchewan Law Review*, 67: 97–135.

Ravelingien, A., and Pennings, G. (2013), 'The right to know your genetic parents: from open-identity gamete donation to routine paternity testing', *American Journal of Bioethics*, 13.5: 33–41.

Report of the Committee of Inquiry into Human Fertilisation and Embryology (1984), hfea.gov.uk/docs/Warnock_Report_of_the_Committee_of_Inquiry_into_Human_Fertilisation_and_Embryology_1984.pdf (last accessed March 2017).

Reynier, P., May-Panloup, P., and Chretien, M. F. et al. (2001), 'Mitochondrial DNA content affects the fertilizability of human oocytes', *Molecular Human Reproduction*, 7.5: 425–9.

Sample, J. (2015), '"Three-parent" babies explained: what are the concerns and are they justified?', *The Guardian*, 2 February, www.theguardian.com/science/2015/feb/02/three-parent-babies-explained (last accessed May 2017).

Shepherd, R., Barnett, J., and Cooper, H. et al. (2007), 'Towards an understanding of British public attitudes concerning human cloning', *Social Science & Medicine*, 65.2: 377–92.

Smajdor, A. (2007), 'The moral imperative for ectogenesis', *Cambridge Quarterly of Healthcare Ethics*, 1.16: 336–45.

Smajdor, A. (2012), 'In defense of ectogenesis', *Cambridge Quarterly of Healthcare Ethics*, 21.1: 90–103.

Smajdor, A., and Cutas, D. (2015), 'Will artificial gametes end infertility?', *Health Care Analysis*, 23.2: 134–47.

Srivastaval, B. (2016), 'Anti-surrogacy bill dashes hopes of aspiring parents', *The Times of India*, 20 September, timesofindia.indiatimes.com/city/ranchi/Anti-surrogacy-bill-dashes-hopes-of-aspiring-parents/articleshow/54418447.cms (last accessed October 2016).

Stacey, J. (2011), Unhitched: Love, Marriage, and Family Values from West Hollywood to Western China, New York: New York University Press.

Szczurek, J. (2013), 'Pursuing the faith: the 616th academic year at the Pontifical University of John Paul II in Krakow', *Analecta Cracoviensia*, 45: 355–90.

Wellcome Trust (1998), 'Public perspectives on human cloning', Input to Human Genetics Advisory Committee and the Human Fertilisation and Embryology Authority's consultation document, Cloning Issues in Reproduction, Science and Medicine.

WHO (1998), 'Ethical, scientific and social implications of cloning in human health', WHA41.10, 1998, www.who.int/ethics/en/WHA51_10.pdf (last accessed February 2017).

Wolf, D. P., Mitalipov, N., and Mitalipov, S. (2015), 'Mitochondrial replacement therapy in reproductive medicine', *Trends in Molecular Medicine*, 21.2: 68–76.

Zegers-Hochschild, F. Adamson, G. D., and de Mouzon, J. et al. (2009), 'International Committee for Monitoring Assisted Reproductive Technology (ICMART) and the World Health Organization (WHO) revised glossary of ART terminology', *Fertility and Sterility*, 92.5: 1520–4.

4

New frontiers in surgery: the case of uterus and penis transplantation[1]

Gennaro Selvaggi and Sean Aas

Various types of organ transplantations are now considered standard procedures: heart and liver transplants lengthen lives; kidney transplants also do so, as well as improving quality of life by reducing or eliminating the need for dialysis. The transplantation of faces and limbs, a more novel set of techniques, improves quality of life without necessarily lengthening or 'saving' lives. An even more recent development is uterus and penis transplantations, which also do not save or lengthen life, but increase reproductive and sexual function and thereby improve quality of life.

This chapter identifies and discusses central ethical issues that are likely to arise in the development of uterus and penis transplantations. These include general issues related to the ethics of surgical research, and specific concerns regarding the rationale of these particular procedures in the context of reproductive and sexual medicine. How should prospective patient-subjects be selected for innovative surgeries? Are these procedures appropriate as treatment for gender dysphoria, or should they be restricted to people whose reproductive and sexual organs have been damaged by illness or accident? Who is most likely to benefit and how are benefit and risks to be judged? What are the alternatives to these transplant surgeries? How should donor organs be sourced? Finally, more broadly, how should we think of these procedures from the perspective of cost-effectiveness – are these expensive, non-life-saving procedures a good use of scarce health resources in light of pressing global needs?

Uterus transplantation and penis transplantation: a brief history

Transplant medicine has made tremendous advances during the last decades. An early milestone was the first successful transplantation of a solid organ, the kidney, in 1956 (Merril et al. 1956; Murray et al. 1960); this was later followed by successful transplantations of other organs, including life-saving transplants of lungs (1963) and hearts (1967) as well as transplants aimed at enhancing quality of life: hand (Dubernard et al. 1999), abdominal wall (Levi et al. 2003), larynx (Birchall et al. 2006), face (Devauchelle et al. 2006), uterus (Brännström et al. 2010; 2014; 2015) and penis (Bateman 2015).

Uterus transplantation in particular raises distinctive issues about the right to reproduce and the relevance of reproductive ability to health and quality of life. The first uterus transplantation was attempted in Germany in 1931, on Lili Elbe, also one of the first trans women to undergo surgery to modify the anatomy of her genitals, removing the male organs (penis and testicles) and constructing female organs (vagina and labia) in order to confirm her desired gender (an early example of what has come to be called 'gender confirmation surgery'). Uterus transplantation was performed together with the reconstruction of the vaginal cavity as a final surgical step. Lili Elbe died three months after this surgery because of infection following rejection of the transplanted uterus. After that early attempt, it took more than eighty years for the first fully successful uterus transplantation (leading to a live birth), performed at the Sahlgrenska University Hospital in Gothenburg, Sweden, by a surgical team directed by the gynaecologist Brännström and the transplantation surgeon Olausson (2015).

Penile reconstructions, which raise overlapping issues, were also originally performed in the 1930s, by the Russian surgeon Nikolaj Bogoras (1936). At the time, surgeons reconstructed the penis using a series of operations, harvesting tissues from other parts of the body (originally skin, fat and muscles, later bones as well), shaping them and transposing them into the pubic area to resemble a natural penis. Penile reconstruction has developed in the intervening years, and is typically offered to two distinct populations: 'cis men' (i.e. individuals whose 'male' gender identity matches the sex and/or gender assigned at birth) whose penis was missing following disease or trauma – say, war injury or botched circumcision; and 'trans men' (i.e. individuals whose 'male' gender identity does not match the 'female' sex and/or gender assigned at birth), some of whom seek anatomical changes to reduce suffering resulting from gender dysphoria. In spite of all the surgical advances and refinements over the past eighty years, an ideal technique for penile reconstruction has not yet been developed.

Surgeons have recently attempted to develop penis transplantation as an alternative to traditional penis reconstruction. The first penis transplantation was performed in China in 2006 on a cis man after traumatic severance of the penis, with the surgery reversed after two weeks due to a negative psychological reaction (Hu et al. 2006). Dr Van der Merwe at Tygerberg Hospital, University of Stellenbosch, in South Africa, performed the first successful penis transplantation in 2014, from a cadaver donor to a male patient who had lost his penis following a failed ritual circumcision; the results have been described as satisfactory, with the patient recovering erectile function, leading to a pregnancy (Bateman 2015). A third operation was performed in 2016 at Boston Massachusetts Hospital, though it is not yet publicly known whether this instance was or will be successful (Caplan et al. 2016).

Uterus and penis transplantations: innovation, care, research

Uterus and penis transplantations are novel surgical procedures, promising significant benefits to those cis and transpeople who are missing an important part of their genital anatomy. However, much remains unknown about the incidence of risks, and the importance of the benefits of these procedures. Going forward, there is an imperative both to offer clinical benefits to prospective subjects and, so far as possible and permissible, to develop generalisable knowledge about these procedures for the sake of future patients. As novel and largely untested therapeutic modalities, uterus and penis transplantations fit into two overlapping ethical categories: clinical care and research.

First and foremost, at present uterus and penis transplantations are sought for clinical benefit; therefore, they are therapies of a particular kind: what are sometimes called *innovative surgeries*. According to Morgenstern (2006), innovative surgery is a novel approach to an unsolved surgical problem or one that offers a substantive improvement of a pre-existing technique. Uterus transplantation provides a novel surgical means to restore or produce fertility, while penis transplantation promises a substantial improvement over existing penile reconstruction techniques. Uterus transplantation restores fertility in cases of absolute uterine factor infertility (AUFI), which is the most significant cause of totally untreatable infertility (Olausson et al. 2014). Currently, there is no alternative that will permit a woman with AUFI to carry on a pregnancy and deliver a live baby. Interestingly, to serve this end a uterus transplantation need not involve permanent implantation of a donor organ. Thus, uterus transplantation is the first ephemeral (i.e. lasting for a short time) transplantation type, whereby the transplanted organ is only temporarily engrafted, and is removed after the baby is delivered (Olausson et al. 2014).

Penis transplantation procedures, as therapies, treat disorders of the genitalia, enhancing quality of life after injury by restoring valued functioning such as urinating while standing and penetrative sexual intercourse. Further, in some trans men, penis transplantation may provide a clinically effective treatment for gender dysphoria, a serious condition that, inadequately treated, poses high risks of both morbidity (depression, anxiety) and mortality (via suicide). In both cis and trans men, penile reconstruction (regardless of the surgical methods used) has a large and usually positive impact on the subject's psychological well-being. Reconstruction reduces organ-specific dysphoria (i.e. 'unhappiness and discomfort') in cis men with genital injury or pathology. In trans men, reconstruction can be an effective treatment for gender dysphoria, particularly in combination with other therapeutic approaches (other surgical procedures like mastectomy, hormonal therapy and psychosocial support) (World Professional Association for Transgender Health 2011). However, there is no consensus in the existing literature regarding the best surgical technique for penile reconstruction (Selvaggi and Elander 2008; Wroblewski et al. 2013).

So far, penile reconstruction has been performed by way of surgical techniques involving various tissues (skin, fat, muscle, etc.) which are taken from one part of the patient's body (forearm, thigh, etc.), shaped into a penis, and repositioned on the pubic area of the same patient. One of the most advanced techniques consists in the use of autologous tissue, the radial forearm flap; however, this usually does not completely fulfil patients' expectations and, as mentioned earlier, it presents a high risk of complications, also leaving the recipient with additional scarring of the donor area (Selvaggi et al. 2006; Selvaggi and Elander 2008; Monstrey et al. 2009; Wroblewski et al. 2013). Figure 4.1 shows the results of a reconstruction with this technique; it clearly demonstrates how the reconstructed penis differs from a typical biological penis; Figure 4.2 shows the residual scar on the donor area. These techniques go under the general name of *phalloplasty*. Specifically for trans men, another surgical alternative is *metaidoioplasty*, which consists in the simple enlargement of the clitoris and possibly the reconstruction of the urethra, to create what is sometimes called a 'micropenis'.

All of these types of penis reconstructions have important limitations, particularly as treatments for gender dysphoria. First, none of these techniques

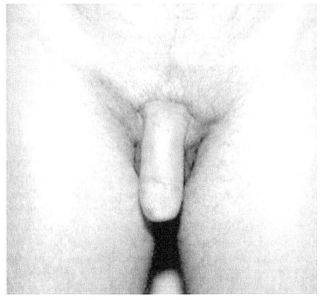

Figure 4.1 Penile reconstruction with radial forearm flap

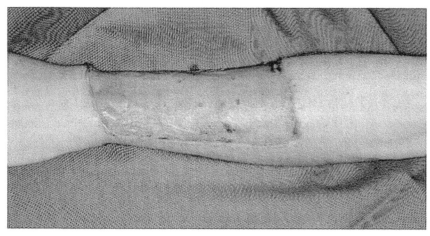

Figure 4.2 Donor-area morbidity (scarring) following the harvesting of a radial forearm flap for penile reconstruction

is able to satisfy entirely patients' expectations: in metaidoioplasty, satisfactory penetrative function is not achievable; in phalloplasty, further surgery is needed to implant a silicone erection device into the reconstructed penis, with additional risks such as infection and extrusion of the erection device. Second, in both cases any observer can recognise that the penis has been reconstructed. Currently, there is no technique that can provide a fully passable penis, and therefore full and seamless functioning in the desired gender identity. Third, other functions such as urinating while standing often require extensive additional 'revision' surgeries, with no guarantee of success. Moreover, in both procedures there is a high risk of complications – the risk of total loss of the reconstructed penis is as high as 2 per cent. In phalloplasty there can be significant donor site morbidity, for example, on the radial forearm. Post-operative recovery typically involves weeks of hospitalisation and many months of rehabilitative therapy.

Due to these limitations, many patients who lack a penis do not attempt reconstruction, instead (in many cases) silently suffering with sometimes severe dysphoria. The existence of this population alone shows that there is a significant health need for improved gender confirmation modalities in trans men, a need which might be met by safe and effective penis transplantation procedures. Nonetheless, so far none of the reported penis transplantation cases have been performed for the purpose of gender confirmation. In spite of some critiques following the first report of penis transplantation in an amputated patient (Dubernard, 2006; Hoebeke et al. 2007), expert reconstructive urologists believe this surgical innovation may offer the best results in penile reconstruction for trans as well as cis men, due to a degree of satisfaction that is currently not available from other methods (Hoebeke et al. 2007). Among all the candidates for penis transplantation, trans men with gender dysphoria represent probably the largest pool. Loss of the penis in cis men due to injury or illness remains relatively rare – the figures are unclear: 7 per cent of military personnel suffer genital injuries in combat, while penile cancer is reported to have a incidence of between 0.3 to 1 in 100,000; however, these figures account for all penile cancers and genital injuries, many of which may not require total penile reconstruction (Bleeker et al. 2008); gender dysphoria in trans men is much more common, with a prevalence ranging from 1:30,400 to 1:200,000 persons in (for instance) Belgium (De Cuypere et al. 2007). Even admitting that not all trans men would want to undergo genital surgery and might be satisfied solely with hormonal therapy and mastectomy, it is still likely that they would represent a large proportion of the patients seeking penile reconstruction by penis transplantation.

Penis and uterus transplantations, therefore, offer substantial benefits to prospective patients. Complex innovations like these should, however, not be pursued lightly, even if the surgeon expects that the patient will benefit. In the case of innovative surgery, particularly surgery which carries significant physical and psychological risks, standard criteria for risk-minimisation include: (1) sufficient scientific background knowledge on the part of the surgical team, probably involving laboratory experiments conducted before procedures are conducted *in vivo*; and (2) stable and sufficient institutional support, with resources reliably available to maximise the chances of a successful procedure, recovery and rehabilitation. Moreover, and relatedly, it is critically important that (3) these innovative surgeries occur within a context of adequate scientific and ethical oversight (Moore 1970; 2000).

It could be argued that this is particularly important in cases such as penis and uterus transplantation, where many innovative procedures will be, though therapeutic, not *merely* therapeutic: the goal will be both to help the patient, and to develop surgical knowledge which can help future patients.

In order to treat someone effectively, it may be necessary to perform these types of surgery; at the end of the day, these types of surgery may be, for some patients, 'the best available treatment'. However, these surgeries need refinement: they are far from perfect at present; the judgement that they are in the patient's clinical interests, therefore, will to some extent reflect a chancy balancing of unknown risks against substantial benefits. But to reduce uncertainty about the risks of these procedures, it is necessary to perform them. This results in a situation that some may regard as uncomfortable: those who receive these types of surgery are to some extent 'human guinea pigs'. Their bodies are utilised to increase knowledge, to collect data and to perfect surgical techniques, in cases where there is, at least, some doubt about whether this is to their clinical benefit. But, at the same time, these 'guinea pigs' are also being treated with what they may regard as the 'best available treatment' at this given time. So, from this perspective any alternative here is morally problematic: either the treatment that the patient may judge as 'the best available treatment' is provided, and the patient may be exposed to unknown risks and be utilised as a means to gather further knowledge and perfect these types of surgery; or the treatment is denied, and a dissatisfactory status quo is preserved.

As new procedures, penis and uterus transplantations will expose patients to unknown and possibly large risks. Mere consent may be enough to justify these risks, but it may not – physicians may well have a duty not to offer excessively risky therapies, even if patients want them. Or it may be that early instances of these procedures are so expensive that some broader future benefit must be cited to support the expenditure. If penis and uterus transplantations are justifiable as clinical research, as well as clinical care, then there may be an additional justification for physicians to perform these dangerous and expensive procedures in cases where the benefits do not definitively and obviously outweigh the risks.

We would do well, then, to consider whether uterus and penis transplanta-tion procedures, in the short term, might be justifiable as research as well as therapy. Framing the issue in this way has a number of advantages, not least the existence of well-worked-out ethical frameworks for evaluating clinical research. Emanuel et al. (2000), for instance, provide an especially perspicacious set of guidelines for evaluating proposed research. Interpreting a variety of national and international research ethics documents, Emanuel et al. argue that medical research, in order to be ethical, must exhibit: *scientific and social value* (evaluation of the treatment, intervention, or theory that will improve health and well-being or increase knowledge); *scientific validity* (use of accepted scientific principles and methods, including statistical techniques, to produce reliable, i.e. reproducible, and valid, i.e. credible, data); *fair subject selection* (selection of subjects so that stigmatised and vulnerable individuals are not targeted for risky research and the rich and socially powerful are not favoured for potentially beneficial research); *favourable risk–benefit ratio* (minimisation of risk, enhancement of potential benefits, risks proportionate to the benefits to the subject and society); *independent review* (review of the design of the research trial, its proposed subject population and risk–benefit ratio by individuals unaffiliated with the research); *informed consent* (provision of information to subjects about the purpose of the research, its procedures, potential risks, benefits and alternatives, so that the individual understands this information and can make a voluntary decision whether to enrol and continue to participate); *respect for enrolled research participants* (permitting withdrawal from research, protecting privacy through confidentiality, informing subjects of newly discovered risks and benefits, informing subjects of results of clinical research, maintaining welfare of the subjects).

These categories, we believe, are useful for evaluating innovative uterus and penis transplantation procedures whether or not these are always 'research' properly speaking (a question we will not attempt to answer here). Among them, social value, favourable risk–benefit ratio and fair subject selection are the most controversial in penis and uterus transplantations; in the ensuing sections, we discuss these in detail.

Social value

If penis and uterus transplantations are justifiable, even in part, because of benefits to future patients, then their justification involves exposing some patients to risk for the benefit of others. It is therefore important that early attempts at these procedures are conducted in a manner that has some chance of producing knowledge that is of real value.

Penis and uterus transplantations raise subtle and difficult issues about the nature of the value that would be produced, if these techniques were developed to the point where they are part of the standard of care for conditions such as penile injury, infertility and gender dysphoria. Penis transplantation as restorative therapy after loss or damage to a cis man's penis should be relatively uncontroversial; restoring injured bodies to physiological normality is a long-recognised and much-pursued goal of medicine. The other cases raise much more complex and controversial issues: in the case of uterus transplantation, the value of and rights to fertility; and in the case of penis transplantation for trans men, the importance of gender identity and the nature of appropriate treatment for gender dysphoria. We discuss the social value of uterus transplantation first, before transitioning to questions about the social value of penis transplantation and its risks and benefits.

Altruism, procreation, parenthood and uterus transplantation

The first ethics guidelines for research and development into uterus transplantation were presented by the International Federation of Gynaecology and Obstetrics in 2009 (Milliez 2009; Olausson et al. 2014): considerations adduced include the value of, and rights to, procreation as well as considerations regarding altruism and alternatives such as adoption. Uterus transplantation, unlike penis transplantation, is intended primarily, perhaps exclusively, to allow a single specific sort of bodily functioning: conceiving, gestating and giving birth to a child. This is evidenced by the fact, mentioned above, that transplanted uteruses are removed after successful birth. Thus, it is important to consider the ethics of this procedure in light of the costs and benefits of the alternatives available to those unable to conceive a child with existing fertility treatments.

In general, there are three options for infertility in the absence of a functioning uterus: (1) choosing not to have children; (2) adopting a child; or (3) pursuing gestational surrogacy. A first point about social value is that adding uterus transplantation to this menu of options can, at the very least, be expected to reduce the incidence of alternative choices. The social value of this eventuality is uncertain, given uncertainty in the magnitude of effects, but it seems likely that widespread availability of uterus transplantation would somewhat decrease the rate of adoptions. Since adoption promotes the well-being of un-parented children, uterus transplantation could be expected to have negative effects in at least this respect, somewhat reducing the supply of prospective adoptive parents. However, the same could be said of any novel fertility treatment; thus, this in itself does not seem to be a decisive objection to uterus transplantation. Moreover, there are serious concerns about the practice of gestational surrogacy, including concerns about exploitation of surrogates (Brännström et al. 2010); any harm done in reducing adoption might be counterbalanced (in whole or in part) by the benefit of reducing the rate of surrogacy and its attendant problems. More recently, Testa and Johannesson (2017) highlight how uterus transplantation offers the mother the possibility of transferring the parental genetic material, and carrying her pregnancy simultaneously.

It might be asked, further, whether restoring fertility – a function that does not seem to be necessary for many people to live fulfilling and happy lives – ought to be a high-priority medical goal in the first place. Of course, many people profoundly value children and family life; however, obviously, bearing children oneself is not the only way to achieve this goal. That said, many people do seem to think that there is a basic human right to procreate (Quigley 2010). Leaving aside, for now, questions about the cost of reproductive assistance relative to other priorities, it does not seem implausible to think that everyone should be able to choose whether to procreate or not. Uterus transplantation promotes this choice in cases where it might otherwise be completely unavailable, thus producing an important kind of social value.

Supposing we grant that there is some general right to procreation, or at least a reason to promote (positive) reproductive freedom, difficult questions arise as to how this right should be realised in people whose gender identity poorly matches the genital anatomy they were born with. One important use of uterus transplantation would be to provide trans women with the opportunity to procreate in a manner reflective of their gender identity.

Sparrow (2008) and Murphy (2014) have discussed whether trans women should be regarded as eligible for uterus transplantation. More specifically, Murphy (2015) considers whether there would be any morally significant reason why trans women, or even cis men, should not be eligible for the same opportunity to gestate as cis women. Noting that other forms of assisted reproduction technology, such as fertility medication, artificial insemination, *in vitro* fertilisation and surrogacy, are regularly offered to both cis women and trans people, Murphy asks why this particular form of assisted reproduction should only be offered for cis women. To be sure, during the early phases of uterus transplantation research, it may make sense to start with cis women, where the procedure is less technically complicated; however, once many of the transplantation issues common to both sexes have been solved, there does not seem to be any specific ethical justification to preclude attempting uterus transplantation in trans women. To do so would, Murphy argues, be a form of unjustified discrimination. Sparrow (2008), who is sceptical of uterus transplantation, focuses his arguments on the allocation of research resources, claiming that all publicly funded research should be limited to sex-typical ways of having children. However, if, as Murphy points out, limiting publicly funded research to 'sex-typical ways' of having children results in discrimination between different groups of patients, it is not clear that cost arguments could suffice. The fact that non-discrimination is costly is not obviously or uncontroversially a strong reason to discriminate. Many believe that justice can sometimes override efficiency, requiring us to devote resources in a manner which sacrifices some overall well-being to promote the possibility of equal treatment of equal persons.

Moreover, as Murphy suggests, Sparrow's proposal seems in some ways to prove too much. After all, the statistically typical form of reproduction for human females involves no artificial fertility aids whatsoever. IVF, for instance, hardly seems 'sex-typical'. Should we, therefore, restrict funding for IVF in favour of an exclusive focus on fertility treatments that attempt to restore biologically 'normal' procreative functioning? That is not at all obvious.

Similar concerns may apply to proposals that would limit uterus transplantation based on the idea that our human right to reproduce only covers reproduction in 'species-typical' ways. The idea here would be that the right to reproduce, though possessed by everyone, is possessed in different ways in 'natural-born men' as opposed to 'natural-born women'. Trans women have a right to reproduce, so that argument goes, as males; thus withholding treatments that might allow them to reproduce in female-typical ways does not violate their reproductive rights.

Again, this argument's focus on species-typicality leaves it unable to justify many existing fertility treatments, which, even in cis women, often involve atypical ways of conceiving and implanting an embryo. Perhaps, however, a proponent of this argument would simply agree that IVF is not included under a human right to reproduce, focusing instead on the right to reproduce as the right to have and make use of a healthy, normal reproductive system. This way of understanding the right is, we suspect, too narrow, fetishising arbitrary facts about statistically typical humans and ignoring the importance of simply being able to do important things, however you do them. We would suggest instead a human right to reproduce in a manner consistent with an individual's basic values and sense of identity; a right, therefore, which would imply that there is strong reason to help trans women reproduce as women.

However, even a more restricted conception of the right to reproduce, a right to reproductive health, could justify uterus transplantation in many trans women. This narrower conception focuses on healthy reproduction as sex-typical reproduction. But health is not just about normality; it is also about the avoidance of suffering. Many trans women suffer from serious cases of gender dysphoria; prior to gender confirmation, they are, therefore, not healthy. Asking them to preserve anatomical structures needed for male-typical fertility until such time as they are ready to bear children is asking them to sacrifice health for reproduction – a demand that ought to be viewed sceptically, on any plausible conception of reproductive rights.

Uterus transplantation as an innovative fertility treatment also opens the possibility of a much more atypical kind of reproduction: uteruses could be transplanted into cis men. This clearly does not solve a health problem; it would, therefore, go beyond the right to reproduce in a healthy way. Even on our broader understanding of the right to reproduce consistent with deeply held beliefs and identities, we doubt that there will often be a forceful claim to uterus transplantation in cis men; it will rarely be the case that a cis man's deeply held values or identity demand that he reproduce by gestating a genetically related child in his own body. Cis men have many other opportunities for socially endorsed, stigma-free parenthood; to most cis men, gestating a child would be at most one option among others. Lacking an option, even if it would be good to have it, is not in itself a rights violation.

That said, one could imagine circumstances under which uterus transplantation in cis men ought to be seriously considered. Murphy (2015) proposes the case of a man who might wish to gestate a child because his female partner is unable to do so (because of uterus incapacity or pregnancy risk to her health) or unwilling to (because she is the main economic provider in the relationship and does not want to lose time to pregnancy). Further, male gestation might appeal to male same-sex couples.

Sparrow worries that, if people identified at birth as male could gestate children, we should be concerned about 'changes in the social meaning and expectations of sexed bodies' (2008: 295). It is, however, hardly obvious that such changes would be a bad thing: feminists have mounted powerful arguments that these meanings and expectations place undue burdens on women (Haslanger 2012). Eventually, offering uterus transplantation in trans women and cis men could provide reproductive options that would be valuable both for the individuals who take advantage of them, and for the structure of the society in which they live. Still, we stress that one need not accept this more radical conclusion to accept a human-rights-based case for uterus transplantation in trans women; that, again, can be understood, in many cases, as a consequence of the right to reproduce free from gender dysphoria.

Risk–benefit ratios in penis and uterus transplantations

There is as yet little detailed discussion of the risks and benefits of penis transplantation in the ethics literature; what there is focuses exclusively on cis men. The group at the Guangzhou General Hospital in China, where the first penis transplantation was performed, attempted to discuss ethical issues related to this innovative surgery, but they did not discuss penis transplantation specifically for trans patients; their analysis was limited to pointing out the importance of considering risk–benefit ratio, informed consent, fair patient selection and psychological assessment, before proceeding with the surgery.

In one of the few extant discussions of the ethics of penis transplantation, Caplan et al. (2016) assert that the development of penis transplantation procedures in live human beings should pause until ethical guidelines are developed. Since, they maintain, penis transplantation is not a life-saving procedure, but rather a high-risk mode for improving quality of life, a precautionary principle requires that procedures should be halted until explicit ethical guidelines are adopted. Ethical discussion must cover the donation of tissue, consent, subject selection, the qualifications of the surgical team and management of both failure and patient dissatisfaction. Caplan et al. conclude that, unless these issues are discussed in more detail, penis transplantation should not be undertaken. We think that Caplan et al. (2016) underestimate the urgency of the clinical needs that penis transplantation meets, and overstate our present state of uncertainty regarding the risks and benefits of the procedure. Caplan is right that only a few anatomical and animal studies on penis transplantation have been reported in the literature (Sonmez et al. 2009; Zhang et al. 2010; Tuffaha, Sacks et al. 2014; Tuffaha, Budihardjo et al. 2014). However, penis reconstruction as currently practised provides a partial model; and here there are many successful and well-studied cases, particularly in trans men (Selvaggi and Elander 2008; Monstrey et al. 2009; World Professional Association for Transgender Health 2011; Wroblewski et al. 2013). It is our judgement that there is sufficient scientific knowledge regarding the underlying anatomical problems to proceed with transplantation attempts in humans, if this is otherwise ethically permissible.

Caplan et al. (2016) adopt a more cautious attitude, in part, because they draw a sharp distinction between penis transplantation as a very risky 'life-enhancing' procedure, which simply improves quality of life, and 'life-saving' procedures, as in many cases of solid organ transplant. More recently, Caplan and Purves (2017) have highlighted how candidates for life-improving organ transplantation may have poor judgement when assessing risks and benefits. This raises the possibility that the decision to opt for surgery may not represent a genuine expression of autonomy. However, as Caplan and Purves note, many other life-improving surgeries widely offered today, as reconstructive (as well as cosmetic) procedures, carry a risk of death from anaesthesia or other complications after surgery such as deadly infections, and yet these procedures are widely offered and chosen. Though it is important for patients to understand risks before undertaking dangerous procedures, banning whole categories of procedure on grounds of concern about misunderstanding of risk goes much further than present practice does or should.

Our view is that all transplants are inherently risky procedures and all carry with them the attendant risks of permanent immunosuppressant therapy (Diaz-Siso et al. 2013; Siemionow 2012). But these risks must be counterbalanced against the benefits, without giving categorical priority to length of life over quality of life. Moreover, it is worth noting that some penis transplantations may extend expected lifespan as well as improve quality of life. For gender dysphoria patients, the degree of dysphoria and subsequent risk of suicide may be so high that adequate gender confirmation might be considered a life-saving procedure.

To be sure, in addition to the risks of immunosuppressant therapy (Siemionow 2012; Diaz-Siso et al. 2013), there are also psychosocial risks, in penis transplantation especially (Caplan et al. 2016). Unlike solid organ grafts (e.g. liver, heart), hand, penis and face transplants are visible, and thus may have more pronounced implications for the recipient's sense of self (Kumnig and Jowsey-Gregoire 2016). This raises the possibility that, as in the Guangzhou case, the recipients would be unable to accept and integrate the graft into their sense of identity and bodily integrity (Carosella and Pradeu 2006). Yet nothing apart from further research in humans could determine whether and to what extent this would actually be the case. In gender dysphoria patients, especially, we believe that post-operative regret will be rare. Significant regret is rare for existing penile construction modalities, even with their many limits. It seems likely that the aesthetic and functional advantages of penis transplantation will produce equivalent or better post-operative satisfaction, though these would have to be weighed against side effects. Still, as Caplan et al. (2016) point out, there will be psychosocial challenges, especially for the early cases, who are likely to receive intense media scrutiny. Caplan et al. are surely right that, *ceteris paribus*, physicians should attempt to minimise these risks as they advise patients about these procedures.

A final issue in risk–benefit evaluation is more fundamental; it concerns the perspective from which risks and benefits should be weighed. One sceptical view on both uterus and penis transplantation might hold that psychosocial benefits, no matter how great, do not justify the enormous risks attendant on complex transplantation surgery and subsequent immunosuppressive therapy. This objection, however, is an anachronism in this anti-paternalistic age of 'patient-centred care'. Patients seek fertility and gender confirmation procedures because a lack of desired reproductive and/ or sexual functioning appears to them, in their own life, as a serious problem. The success or failure of these procedures, particularly in trans populations, is measured primarily, if not exclusively, by self-reported patient satisfaction (Kuiper and Cohen-Kettenis 1988). Clinicians do not and should not decide how big a risk justifies these benefits. They should, of course, advise as to the nature of the risks; when these are high, as in uterus and penis transplantations, they should take extra steps to ensure that the patient understands them. The same goes, we argue, for surgeons as clinician-researchers, in the development of uterus and penis transplantations. Given the highly personal nature of the goods at stake in uterus and penis transplantations, we should take the fact that well-informed, competent patients are willing to undergo these experimental procedures as reason enough to believe that they have a sufficiently favourable risk–benefit balance.

Fair subject selection: cadaver donors and living donors for uterus and penis transplantations

Uterus and penis transplantations also raise issues of 'fair subject selection'. That physicians should largely defer risk–benefit judgements to well-informed and competent patients does not mean that many or all prospective patients will be competent and well informed. Moreover, given the risks of this procedure, there may be some limits to deference: patients who are not sure that they themselves will benefit from the procedure, but wish to assist research for the sake of others, might reasonably be refused, given the gravity of the decision to undertake a lifetime of immunosuppressant therapy. Deference to patient judgement on risk and benefit does not entail a totally 'hands-off' approach to patient decision. Physicians retain some discretion in restricting their own participation in innovative procedures to cases where the patient can actually expect some benefit. As Caplan et al. (2016) argue, surgeons would be well advised to encourage patients who do understand the distinctive risks of the procedure and who are most likely to benefit. Apart from the physiological and anatomical requirements, personal factors relevant to likely penis transplantation success in cis men might include family and social support, lack of surgical alternatives which could adequately fulfil subject expectations, and strong perceived dysphoria in the preoperative state.

Uterus as well as penis transplantations, as transplant procedures, involve two patients: a donor as well as a recipient. There are also substantial ethical issues about the 'sourcing' of uteruses and penises for transplantation. In the uterus transplantation procedures performed so far, surgeons have opted for living donors. The use of a living donor increases the likelihood of success due to higher chance of compatibility (e.g. donation from a relative of the recipient) and ease of procurement; surgery from a living donor can be planned in advance, obviating problems of emergency extraction and transplantation that come along with opportunistic cadaveric donation. Sourcing organs from living donors makes sense in uterus transplantation. Uteruses, unlike many other organs, are sometimes truly 'surplus' to the needs of living donors. Some women have no further plans for pregnancy, while others are unable to conceive and gestate a child despite having a healthy uterus. Uterus donation has a risk level similar to hysterectomy, now a standard procedure (Olausson et al. 2014). Still, recent advances may make dead donor uterus retrieval more feasible in the near future (Testa et al. 2017).

For the cases performed to date of penis transplantation in cis men, surgeons have opted instead for cadaveric donors. It is likely that people would have different attitudes towards the donation of symbolically important organs such as the penis or face, as compared to 'internal' organs such as the kidneys, liver and uterus. As Caplan et al. (2016) point out, 'blanket' consent to donation, whether actual ('opt-in') or presumed ('opt-out'), may not suffice in these cases. 'Opt-out' consent presents particular challenges; it seems implausible to 'presume' that people who have never considered organ donation at all would consent to donating such a symbolically important item. Moreover, in both systems, in addition to the deceased person's wishes, the interests and values of their surviving family should perhaps also be considered (Dickenson and Widdershoven 2001; Roels and Rahmel 2011).

Indeed, it is, at first glance, not entirely clear that the practice of sourcing solely cadaveric tissue for penis transplantation should continue; living donation, if sources could be found, would have the advantage of specific informed consent, along with the other practical advantages discussed above regarding uterus transplantation. It might be thought that living donation would be unlikely, given a dearth of 'surplus' penises. This is probably correct, though existing sources should be explored. A small number of cis men suffer from body integrity identity disorder with respect to their penises; if voluntary amputation is permissible in some such cases, this could be one source of living donors (Favazza 1989; Dua 2010). Another, more numerous population (mentioned by the Guangzhou team) would be trans women who seek to be relieved of their male-identified genital anatomy. As it turns out, current gender affirmation surgery modalities for trans women (i.e. vagino-clitoro-labioplasty) require parts of the penis (skin envelope, glans, skin from the scrotum) for the construction of female genitalia (Selvaggi and Bellringer 2011). However, if gender affirmation surgery is developed to the point that penile elements are no longer required to construct the female genitalia, then trans women patients might become a donor source for penises for penis transplantation patients – possibly a particularly apt and psychologically satisfying one for penis transplantation patients seeking treatment for gender dysphoria.

For now, then, it seems unlikely that there will be a significant supply of living penis donors. Cadaveric donors may, therefore, be the only option for penis transplantation procedures in the short term. Though there may be certain practical difficulties in storing and transporting tissues, these do not seem to be decisive, given that all three (somatically) successful penis transplantation operations have used cadaveric donors. Further, given that the only living donors presently available will be patients suffering from body integrity identity disorder, use of living donors would embroil penis transplantation surgeons in raging controversies about the permissibility of voluntary medical amputation. Thus, for now, cadaveric donation, perhaps with specific consent, appears preferable, while avenues for increasing the quantity and quality of tissue sources for the future are explored.

Uterus and penis transplantation: scarce resources

There is, we believe, a strong case that surgeons would do no wrong to patient-subjects or potential donors when offering and performing uterus and penis transplantations as innovative research procedures. This, however, does not determine whether the development of these procedures is good or just overall. Complex transplant procedures such as uterus and penis transplantations are expensive. Caplan et al. (2016) suggest the costs for penis transplantation could run into the millions; uterus transplantation is very expensive as well, particularly given that, as 'ephemeral' transplants, they involve three major procedures – removal from donor, implantation, and later removal from recipient. These procedures, we have argued, could have significant medical benefits in clinical practice. But that is not enough in itself to justify them; these benefits have to be worthwhile, given the costs (Wilkinson and Williams 2016). Resources are scarce, for both research and care, and there are many people in the world who need medical help (Persad et al. 2009). This is true both within health systems that may need to decide whether to cover uterus and penis transplantation procedures, and globally, where large portions of the overall burden of disease are attributable to conditions that can be treated much more cheaply. Moreover, insofar as uterus and penis transplantations are primarily (if not exclusively) life-enhancing rather than life-saving, we might think that they should be deprioritised relative to other expensive transplant treatments, which actually save lives.

It is, in fact, difficult to derive estimates of the likely cost-effectiveness of uterus and penis transplantations, either for cis or trans people. This is in part because the benefits of these procedures are, though probably very real, difficult to quantify and compare to the benefits of other interventions that might be researched or deployed using the same resources. There is in particular little if any systematic work that attempts to quantify the disease burden of gender dysphoria or the cost-effectiveness of various gender affirmation surgery modalities. There is more work on the cost-effectiveness of fertility treatments, though it does not discuss uterus transplantation and covers only cis women (Devlin and Parkin 2003). Still, we would stress a point, presented earlier, that insofar as uterus and/or penis transplantation may constitute the only effective treatment for some severe cases of gender dysphoria, these procedures are likely to have a significant positive effect on both morbidity and mortality in this population, which exhibits high rates of psychiatric co-morbidity and suicide. Again, the issue is not simply life enhancement vs. life preservation; treatments for gender dysphoria accomplish both.

There is normative uncertainty here, as well. Proponents of cost-effectiveness analysis in global health policy often presuppose a certain sort of 'consequentialist' framework, in which relieving a given quanta of health burdens is equally important whoever experiences that burden (Persad et al. 2009). This view puts a strong onus of proof on proponents of any expensive treatment, since the relevant resources can instead be devoted to inexpensive and highly effective interventions in the developing world (e.g. malaria prevention). However, open questions remain about the right of national health systems to prioritise their own citizens over others in the world who might make better use of national health resources. Moreover, even focusing on a given health system, advocates for patients with difficult-to-treat conditions argue that these patient populations should not be abandoned entirely; efforts should be made to develop adequate treatments for every serious health condition, even if some of these conditions are much more expensive to treat than others (Gericke et al. 2005). If any of these arguments are correct, then advocates of uterus and penis transplantations need not show that they are cost-effective relative to, say, malaria prevention in the developing world, or even that they are not much more expensive than treatments offered as standard in developed-world health systems. It may be enough to show that, without access to uterus and penis transplantations, some people – especially some transpeople – will not have an adequate chance at a decent life.

Conclusions

Both uterus and penis transplantation research are ethically plausible when regulated by an ethical framework, though of course many hard questions remain. The benefits of these procedures might include treatment for gender dysphoria in patients who do not see adequate benefits in existing gender affirmation surgery modalities; restoration of normal function in cis populations who have lost penises to injury; and provision of gestational ability in both trans and cis women. Further research is required, however, to understand the magnitude of these benefits and the nature of patients' normative claims to them. The risks of these procedures are substantial, but patient interest suggests that trans and cis people, as competent evaluators of their own interests, reasonably believe that the benefits justify the risks. Further research will provide more information on the risks of these procedures relative to alternatives, allowing patients to make informed decisions. Organ sourcing raises challenges in both uterus and (especially) penis transplantations, but these do not seem to be insurmountable. Finally, given the global burden of disease, it is important to carefully consider whether uterus and penis transplantations, as innovative surgery or (eventually) as routine care, are a good use of scarce health resources. Much more research is required here, as well, to understand the cost-effectiveness of these procedures; still, we doubt that cost considerations will or should halt this line of inquiry altogether.

Appendix: terminology

- Absolute uterine factor infertility (AUFI): medical condition in which the uterus is absent or diseased, and therefore does not allow normal embryonic implantation.
- Cis man/woman: man/woman whose gender identity matches the sex and/or gender assigned at birth.
- Gender affirmation confirmation surgery: refer to all types of surgery aimed to align the genital anatomy to the gender identity. This includes chest-contouring surgery (e.g. mastectomy in trans men; breast augmentation in trans women), genital surgery (penile reconstruction, or phalloplasty, for trans men; vaginoplasty, clitoro-labioplasty, orchidectomy for trans women).
- Gender dysphoria: distress experienced by a person as a result of the sex and gender assigned at birth.
- Trans man/woman: man/woman whose gender identity does not match the sex and gender assigned at birth.
- Vascularised composite allotransplantation (VCA): also referred as composite tissue allotransplantation, the term used to refer to transplantation of an organ composed of several kinds of tissues (i.e. skin, muscle, bone) such as face, hand, arm and penis.

Note

1 Thanks are due to Simona Giordano for her continuous support and suggestions while we were completing this chapter.

References

Bateman, C. (2015), 'World's first successful penis transplant at Tygerberg Hospital', *South African Medical Journal*, 105.4: 251–2.

Birchall, M. A., Lorenz, R. R., Berke, G. S., Genden E. M., Haughey, B. H., Siemionov, M., and Strome, M. (2006), 'Laryngeal transplantation in 2005: a review', *American Journal of Transplantation*, 6.1: 20–6.

Bleeker, M. C., Heideman D. A., Snijders P. J., Horenblas, S., Dilner, J., and Meijer, C.J. (2008), 'Penile cancer: epidemiology, pathogenesis and prevention', *World Journal of Urology*, 27.2: 141–50.

Bogoras, N. (1936), 'Über die volle plastische Wiederherstellung eines zum Koitus fähigen Penis (Peniplastica totalis)', *Zentralblatt für Chirurgi*, 22.1: 1271–6.

Brännström, M., Johannesson L., and Bokstrom H. et al. (2015), 'Livebirth after uterus transplantation', *The Lancet*, 385.9968: 607–16.

Brännström, M., Johannesson L., and Dahm-Kahler P. et al. (2014), 'First clinical uterus transplantation trial: a six month report', *Fertility and Sterility*, 101.5: 1228–36.

Brännström, M., Wranning C. A., and Altchek, A. (2010), 'Experimental uterus transplantation', *Human Reproduction Update*, 16.3: 329–45.

Caplan, A. L., Kimberly, L. L., Parent, B., Sosin, M., and Rodrigues, E. D. et al. (2016), 'The ethics of penile transplantation: preliminary recommendations', *Transplantation*, 101.6: 1200–5.

Caplan, A., and Purves, D. (2017), 'A quiet revolution in organ transplant ethics', *Journal of Medical Ethics*, 43.11: 797–800.

Carosella, E. D., and Pradeu, T. (2006), 'Transplantation and identity: a dangerous split?', *The Lancet*, 368.9531: 183–4.

De Cuypere, G., Van Hemelrijck, M., and Michel, M. A. et al. (2007), 'Prevalence and demography of transsexualism in Belgium', *European Psychiatry*, 22.3: 137–41.

Devauchelle, B., Badet, L., and Lengele, B. et al. (2006), 'First human face allograft: early report', *The Lancet*, 368.9531: 203–9.

Devlin, N., and Parkin, D. (2003), 'Funding fertility: issues in the allocation and distribution of resources to assisted reproduction technologies', *Human Fertility*, 6.2: S2–S6.

Diaz-Siso, J. R. (2013), 'Vascularized composite tissue allotransplantation – state of the art', *Clinical Transplantation*, 27.3: 330–7.

Dickenson, D., and Widdershoven, G. (2001), 'Ethical issues in limbs transplants', *Bioethics*, 15.2: 110–24.

Dua, A. (2010), 'Apotemnophilia: ethical considerations of amputating a healthy limb', *Journal of Medical Ethics*, 36.2: 75–8.

Dubernard, J. M. (2006), 'Penile transplantation?', *European Urology*, 50.4: 664–5.

Dubernard, J. M., Owen, E., and Herzberg, G. et al. (1999), 'Human hand allograft: report on first 6 months', *The Lancet*, 353.9161: 1315–20.

Emanuel, E. J., Wendler, D., and Grady, C. (2000), 'What makes clinical research ethical?', *Journal of the American Medical Association*, 283.20: 2701–11.

Favazza, A. R. (1989), 'Why patients mutilate themselves', *Hospital Community Psychiatry*, 40.2: 137–45.

Gericke, C. A., Riesberg, A., and Busse, R. (2005), 'Ethical issues in funding orphan drug research and development', *Journal of Medical Ethics*, 31.3: 164–8

Haslanger, S. (2012), *Resisting Reality*, Oxford: Oxford University Press.

Hoebeke, P. (2007), 'Re: Weilie Hu, Jun Lu, Lichao Zhang. et al. A preliminary report of penile transplantation Eur urol 2006;50:851–3', *European Urology*, 51.4: 1146–7.

Hu, W., Lu, J., and Zhang, L. et al. (2006), 'A preliminary report of penile transplantation', *European Urology*, 50.4: 851–3.

Kuiper, B., and Cohen-Kettenis P. (1988), 'Sex reassignment surgery: a study of 141 Dutch transsexuals', *Archives of Sexual Behaviour*, 17.5: 439–57.

Kumnig, M., and Jowsey-Gregoire, S. G. (2016), 'Key psychosocial challenges in vascularized composite allotransplantation', *World Journal of Transplantation*, 6.1: 91–102.

Levi, D. M., Tzakis, A. G., and Kato T. et al. (2003), 'Transplantation of the abdominal wall', *The Lancet*, 361.9376: 2173–6.

Merril, J. O., Murray, J. E., and Harrison, J. H. et al. (1956), 'Successful homotransplantations of the human kidney between identical twins', *Journal of the American Medical Association*, 160.4: 277–82.

Milliez, J. (2009), 'Uterine transplantation', *International Journal of Gynecology and Obstetrics*, 106.3: 270.

Moore, F. D. (1970), 'Therapeutic innovation: ethical boundaries in the initial clinical trials of new drugs and surgical procedures', *CA A Cancer Journal for Clinicians*, 20.4: 212–27.

Moore, F. D. (2000), 'Ethical problems special to surgery: surgical teaching, surgical innovation, and the surgeons in managed care', *Archives of Surgery*, 135.1: 14–16.

Monstrey, S., Hoebeke, P., and Selvaggi, G. et al. (2009), 'Penile reconstruction: is the radial forearm flap really the standard technique?', *Plastic and Reconstructive Surgery*, 124.2: 510–18.

Morgenstern, L. (2006), 'Innovative surgery's dilemma', *Surgical Innovation*, 13.1: 73–4.

Murphy, T. F. (2014), 'Assisted gestation and transgender women', *Bioethics*, 29.6: 389–97.

Murray, J. E., Merrill, J. P., and Dammin, G. J. et al. (1960), 'Study of transplantation immunity after total body irradiation: clinical and experimental investigation', *Surgery*, 48.1: 272–82.

Olausson, M., Johannesson, L., and Brattgard, D. et al. (2014), 'Ethics of uterus transplantation with live donors', *Fertility and Sterility*, 102.1: 40–3.

Persad, G., Wertheimer, A., and Emanuel, E. J. (2009), 'Principles of allocation for scarce medical resources', *The Lancet*, 373.9661: 423–31.

Quigley, M. (2010), 'A right to reproduce', *Bioethics*, 24.8: 403–11.

Roels, L., and Rahmel, A. (2011), 'The European experience', *Transplant International*, 24.4: 350–67.

Selvaggi, G., and Bellringer, J. (2011), 'Gender reassignment surgery: an overview', *Nature Reviews Urology*, 8.5: 274–82.

Selvaggi, G., and Elander, A. (2008), 'Penile reconstruction/formation', *Current Opinion in Urology*, 18.6: 589–97.

Selvaggi, G., Monstrey, S., and Hoebeke, P. et al. (2006), 'The donor site morbidity of the radial forearm free flap after 125 phalloplasties in gender identity disorder', *Plastic Reconstructive Surgery*, 118.5: 1171–7.

Siemionow, M. (2012), 'Impact of reconstructive transplantation on the future of plastic and reconstructive surgery', *Clinics in Plastic Surgery*, 39.4: 425–34.

Sonmez, E., Nasir, S., and Siemionow, M. Z. (2009), 'Penis allotransplantation model in the rat', *Annals of Plastic Surgery*, 62.3: 304–10.

Sparrow, R. (2008), 'Is it "every" man's right to have babies if he wants them? Male pregnancy and the limit of reproductive liberty', *Kennedy Institute of Ethics Journal*, 18.3: 275–99.

Testa, G., Anthony, T., and McKenna, G. J. et al. (2017), 'Deceased donor uterus retrieval: a novel technique and workflow', *American Journal of Transplantation*, doi:10.1111/ajt.14476.

Testa, G., and Johannesson, L. (2017), 'The ethical challenges of uterus transplantation', *Current Opinion Organ Transplantation*, 22.6: 593–7.

Tuffaha, S. H., Budihardjo, J. D., Sarhane, K. A., Azoury, S. C., and Redett, R. J. (2014), 'Expect skin necrosis following penile replantation', *Plastic and Reconstructive Surgery*, 134.6: 1000e–1004e.

Tuffaha, S. H., Sacks, J. M., and Shores, J. T. et al. (2014), 'Using the dorsal, cavernosal, and external pudendal arteries for penile transplantation: technical considerations and perfusion territories', *Plastic and Reconstructive Surgery*, 134.1: 111e–119e.

Wilkinson, S., and Williams, N. J. (2016), 'Should uterus transplants be publicly funded?', *Journal of Medical Ethics*, 42.9: 559–65.

World Professional Association for Transgender Health (2011), *Standards of Care for the Health of Transsexual, Transgender, and Gender Nonconforming People*, 7th version, www.wpath.org (last accessed 1 July 2016).

Wroblewski, P., Gustafsson, J., and Selvaggi, G. (2013), 'Sex reassignment surgery for transsexuals', *Current Opinion in Endocrinology & Diabetes and Obesity*, 20.6: 570–4.

Zhang, L. C., Zhao, Y. B., and Hu, W. L. (2010), 'Ethical issues in penile transplantation', *Asian Journal of Andrology*, 12.6: 795–800.

5

Freedom, law, politics, genes: the case of mitochondrial transfer

Iain Brassington

In early 2015, the UK became the first country to make explicit legal provision for the use of mitochondrial transfer techniques leading to a live human birth. Mitochondrial transfer offers a means to prevent mitochondrial illnesses being passed from a mother to her children, as they would be inevitably without the process. Two methods are possible: maternal spindle transfer, and pronuclear transfer. In both, nuclear material is removed from a cell that has faulty mitochondria, and inserted into an enucleated cell with healthy mitochondria; the difference boils down to one of whether that nuclear material is taken from an unfertilised or fertilised ovum. In this chapter, I shall examine some of the senses in which mitochondrial transfer, and the law's handling of it, might be taken to relate to scientific freedom. I shall try to avoid taking a position; my concern is simply to look at some of the potential argumentative fault lines. For the sake of ease, I shall conflate the terms 'maternal spindle transfer' and 'pronuclear transfer' under the term 'mtDNA transfer'.

The purported advantages of mtDNA transfer are easily explained. Mitochondria are mainly concerned with energy production within the cell, and so any faults in them lead to the cell as a whole not functioning as it ought. This malfunction can manifest as any of a range of conditions – some mild, some serious, some fatal – and the term 'mitochondrial disease' is a basket term that captures this range (Tuppen 2010). These conditions may present differently throughout a person's life. The ability to 'edit' a person's mitochondrial inheritance at the very earliest stages of life, or even prior to their life beginning, provides an opportunity to improve what might otherwise be a significantly diminished quality of life. (For the sake of what follows, I shall leave to one side the debates surrounding the non-identity problem: there is so little mtDNA compared to nuclear DNA within a cell that I shall assume that mtDNA transfer alters a determinate person's life, not the person who lives it. If I am wrong on this, and the identity of the child born changes according to whether or not mtDNA transfer has been used, it would mean that no child is 'treated' by the technique, so much as replaced by an overwhelmingly similar sibling; but for the sake of my argument here, I think the point would be nugatory) (Wrigley et al. 2015; Liao 2017; Rulli 2017).)

In short, mtDNA transfer is a form of genetic engineering that promises to make future lives appreciably better. Moreover, because mitochondria are inherited directly down the maternal line, the ability also means that mitochondrial faults that might otherwise pass down the generations can be prevented in subsequent generations. As we shall see, this does raise a couple of lines of concern with respect of the ethics of mtDNA transfer.

Freedom from legal restriction

The background to the UK's regulation of mtDNA transfer goes back to the Human Fertilisation and Embryology Act (1990), which was modified by a further Act in 2008; in effect, the mtDNA transfer regulations constitute a gloss on that modificatory legislation.

The 1990 law's attitude to genetic manipulation was fairly straightforward: it stipulated in section 3(1) that '[n]o person shall bring about the creation of an embryo except in pursuance of a licence', and in section 3(2) it stipulated that '[n]o person shall place in a woman (a) a live embryo other than a human embryo, or (b) any live gametes other than human gametes' (where, *per* section 1(1)(b), 'embryo' was taken to include 'an egg in the process of fertilisation'). On the face of it, this would have permitted mtDNA transfer, since embryos manipulated in this way are clearly human embryos. However, section 3(3)(d) said that a licence to bring about the creation of an embryo could not authorise 'replacing a nucleus of a cell of an embryo with a nucleus taken from a cell of any person, embryo or subsequent development of an embryo'. No reference was made to the purpose for which nuclear material may be replaced. mtDNA transfer involves the nuclear matter from one cell being implanted into an enucleated second cell. Therefore section 3(2) would have been moot: there is no need to worry about permitting the implantation of a cell the creation of which is not permitted.

Yet there could be situations in which we would want to replace a cell nucleus for reasons other than reproduction – say, as part of a research programme; the 2008 Act modified the law such as to allow such manipulations. Subsection 3(3)(d) was to be deleted, and subsection 3(2) changed to specify that only 'permitted' embryos could be implanted in a woman. Section 3(5) of the 2008 Act added to the 1990 Act a new section, 3ZA, to follow section 3. This new section defined 'permitted' embryos as being those created in the fertilisation of a permitted egg by permitted sperm – in turn, defined as eggs and sperm the nuclear or mitochondrial DNA of which has not been altered – and that had itself undergone neither alteration of its own nuclear or mitochondrial DNA, or had any cells added to it except by regular cell division. Thus nuclear matter could be transferred from one embryo to another, so long as the process did not result in gestation.

Importantly, though, section 3ZA(5) made one exception to this rule: regulations may provide that

- an egg can be a permitted egg, or
- an embryo can be a permitted embryo, even though the egg or embryo has had applied to it in prescribed circumstances a prescribed process designed to prevent the transmission of serious mitochondrial disease.

In other words, the door was left open to permission being given for mitochondrial transfer techniques to be used for the sake of avoiding mitochondrial disease. What happened in 2015 was simply that a regulation was approved that provided for, and set out the conditions of, this eventuality (Human Fertilisation and Embryology (Mitochondrial Donation) Regulation 2015).

Though the UK's regulations on what can be done to embryos remain fairly tight – no other form of genetic engineering that would generate an embryo to be implanted is currently permitted – there is a freedom granted by the prevailing regulations in the UK that did not exist before, and that does not exist in a number of other countries (Hamzelou 2016).[1]

This legislative liberalisation has not been without controversy. Two significant lines of objection are worth mentioning. The first is that the technique may be unsafe, and that freeing us to pursue it may not be desirable for welfarist reasons. The second is a version of the slippery slope argument – that there are some freedoms that we ought not to countenance, and that genetic modification of humans may be one of them. Such a line of argument need not appeal to the welfare of particular people. I shall suggest that both these lines fail. On the other side of the debate, there is a question about whether pursuing such technologies may be freedom-enhancing, therefore not only the kind of thing that liberal states should tolerate, but the kind of thing that they ought to endorse. This line of argument is not without its own complexities, as we shall see.

Freedom from risk

The safety concern is straightforward: that genetic engineering is, in the grand scheme of things, a very young science, and that such is the import of our genomes to our well-being that we cannot be certain that we would not be generating much bigger problems than we were hoping to solve. Faulty mitochondria are passed down maternally, to all of any affected woman's children; they cannot be passed on to her grandchildren if she has no daughters. By the same token, should interference with natural mtDNA inheritance unexpectedly create its own problems, we might expect them to be passed down matrilineally as well. We may be condemning not just one person, but an indefinite number of persons over the subsequent generations, to suffer the consequences of a misguided intervention.

Concerns along these lines were important factors in the debates on the regulations in the House of Commons in 2014 and 2015. Some MPs made a case for a precautionary approach to mtDNA transfer, which would presumably mean deferring the new licences – perhaps indefinitely. The fact that the alterations would be heritable was a cause for concern for others; thus Robert Flello MP pointed out that 'the research that has been done talks about generations of mice or of monkeys, but that does not address the fact that until there have been three, four, five or 10 generations, we will not know what the long-term effects are' (Flello 2014).

The safety of an action is a legitimate factor in decisions about whether to perform or allow it. The moral gravity of the freedom to make use of mtDNA transfer techniques does depend fairly straightforwardly on how safe they are. If we are Millian by inclination, a version of the harm principle would suffice to carry this argument: if there is nothing improper about curtailing someone's liberty to prevent harm to others (Mill 1985; Ellison 2015), there is presumably also nothing improper about not granting that liberty to prevent harm to others who may not exist yet, but the quality of whose existence will be influenced by the activity in question. If we are non-Millian utilitarians, without the same overriding concern to preserve liberty whenever possible, the justification of the action will depend much more straightforwardly on the expected welfare return – and the higher the risk of diminished welfare, the weaker the justification. For non-consequentialists, it is likely still to be possible to construct an argument along the lines that a virtuous or dutiful person would be wary of passing a law that permits something with a significant risk of harm, even if an appeal to harm is not the main philosophical driver. Whatever our moral position, the greater the risk of harm, the weaker the justification for making the regulations more permissive. It is not unreasonable to think that one of the primary duties of the state is to afford as much protection as possible to citizens and future citizens from risks that they might otherwise face.

Speaking in the 2015 Commons debate, Jane Ellison MP attempted to answer these concerns, drawing MPs' attention to the fact that

> [t]here has been much discussion of the safety of mitochondrial donation techniques. As I have said, three reports have been produced by the HFEA-convened expert panel during the current Parliament. On each occasion, the panel has concluded that there is nothing to indicate that the two donation techniques are unsafe. Although the panel has recommended that further experiments should be conducted, it expects such research to support the conclusions that it has reached so far. (2015)

This kind of statement is probably not enough to satisfy the precautionary critic of mtDNA transfer techniques. Finding nothing to indicate that the they are unsafe is compatible with not having done any research on the matter at all: that, too, would allow us to say that there is no indication of any danger. Of course, that point is hyperbolic; but the point would remain that even when research has been done, the absence of evidence of danger is not evidence of absence. In this light, the critic is likely to see the claim that a panel expects future results to support current conclusions as scant reassurance when the unexpected results may be catastrophic, and persist for several generations. It's the unexpected consequences that should worry us most, they will argue – and sensibly. The case for liberalising the regulations is not hereby successfully defended.

Yet the thing that is missed by the precautionary account is not so much that mtDNA transfer is risky, but that there is also a risk in *not* pursuing it. It is worth repeating that mitochondria are inherited directly down the maternal line, which means that a woman who is a carrier of faulty DNA will pass it on to any offspring she has; if she has daughters, they will pass it on to their children. Therefore, the decision to be taken is not one between a risky biotechnological procedure and nothing; it's between what I shall allow for the sake of the argument is a risky biotechnological procedure, and a risky natural conception. The precautionary argument against liberalising the regulations is weakened when we consider that the counterfactual is not necessarily risk-minimising. Indeed, while it is true that the absence of evidence for harm is not evidence of absence, we have a pretty good reason to think that *not* pursuing mtDNA transfer is at best risk-stabilising; and, given that the choice is between keeping the risk level stable and intervening in a way that we believe will reduce it, there would seem to be a good moral argument available in favour of the procedure. Taking this line allows the supporter of liberalisation to turn the claim about the role of the state in protecting citizens from risk on its head.

The question then becomes one not so much of whether we should liberalise the regulations, as about what a reasonable level of certainty concerning the safety of the procedure is. Yet how we should establish what is reasonable is a puzzle in itself, there being no obvious rubric for settling the matter. One might be tempted to appeal to an ideal of open debate, the participants in which gradually home in on the most acceptable answer. When it comes to deciding what the law should be, this kind of procedure has a certain prima facie democratic appeal. However, in the context of concerns about risk, the problem is that data to inform this debate may be hard to come by, even assuming that the people charged with the task of framing the laws go about the task disinterestedly. Until mtDNA transfer is performed in a statistically significant number of humans, we won't be able to say a great deal with certainty about how it's likely to affect them; and even the course of the mitochondrial diseases we are aiming to prevent isn't always predictable. Nevertheless, a precautionist position is still quite weak if all it does is insist that there might be some undesirable outcome that is worse than doing nothing: it's true that more or less anything *might* happen, but it doesn't follow that we need to lose sleep over every logically possible sequela. What is needed is a plausible account of what the undesirable consequences might be, or even just the mechanism by which they'll arise. Absent that, there is no obvious reason to take it all that seriously.

Freedom from restraint

The second line of argument turns on the claim that mtDNA transfer amounts to human genetic engineering – which, inasmuch as that it involves deliberately manipulating an aspect of the human genome for a specific purpose, is accurate. Once we have accepted the principle of human genetic engineering, is there any clear reason not to accept it in any and all other circumstances? For sure, implanting a genetically altered embryo to avoid mitochondrial disease requires a specific legal provision; but it is easy enough to make provision for more or less anything, and the mtDNA provision is important because it appears to break the back of the principled prohibition on human genetic engineering. From a certain perspective, there is little difference between adding desirable genes to a zygote or embryo by mtDNA transfer and adding other desirable genes by other methods.

The spectre of eugenics haunts the debate at this juncture, and the word has been used to describe mitochondrial transfer in the media (Newman 2013). On the face of it, there may be a decent reason for this. A short book published in 1929 by Leonard Darwin explains what he means by the term: eugenics, for him, promises 'wonderful effects' for the human race, with the object of 'improv[ing] the breed of the whole nation' (1929: 26). In particular, Darwin states that

> [w]e can at all events assert that there are many kinds of men that we do not want. These include the criminal, the insane, the imbecile, the feeble in mind, the diseased at birth, the deformed, the deaf, the blind, etc., etc. . . . When eugenics comes to be more studied, it will be possible to give advice with greater confidence than at present . . . Even with our present knowledge it is, however, unquestionable that great benefits might be conferred on future generations by the voluntary renunciation of parenthood by the diseased and by such as are very likely to be the carriers of the hidden seeds of disease. (1929: 25, 33)

It takes little imagination to identify faulty mitochondria as one of the hidden seeds of disease, and women with faulty mitochondria as carriers.

Renunciation of parenthood has always been an option for people identified as carriers of congenital illness; in 2009, while considering mitochondrial illness and possible responses to being a carrier, Ruth Chadwick (2009) observed that '[o]ne option is to avoid reproduction altogether, where the risk is perceived to be too great; another is to decide to proceed nevertheless'. mtDNA transfer provides a third option. However, the general thrust remains the same: what is under consideration is a set of methods by which we can avoid bringing to birth individuals who belong to one of the 'many kinds of men that we do not want'. In short, it seems as though mtDNA transfer can be characterised as eugenic. Darwin's criteria for desirability may strike us as odd or naive – though he is reluctant to set out precise standards that we ought to try to meet in our eugenic programmes, he still feels confident to say that '[t]he most practical way of judging grown men is by seeing how they are fulfilling the duties of the positions which they actually hold' (1929: 26) – but we can still say that there are certain inherited characteristics that it is better to be without by any reasonable standard of judgement (Brassington 2013: ch. 1), and all this could be fed into what looks like a eugenicist account or norm. Yet eugenics is the kind of thing that attracts a great deal of moral suspicion from many quarters. Would liberalisation of the law on mitochondrial transfer perhaps represent the preamble a eugenicists' charter? This may be a freedom that we ought not to grant.

I think that this is a worry that can be answered, though we have to untangle its elements. Let's allow for the moment that we are right to look at eugenics with suspicion. Even so, though some action may be the kind of thing that a eugenicist would endorse, it does not follow that everyone who endorses it is a eugenicist. It might be that mtDNA transfer is eugenicist; but the fact that we can construct a eugenicist argument in favour of it won't show that: we'd need to be able to show that there could be no non-eugenicist argument for it.

Admittedly, though, if eugenics is about improving the stock of humanity, then it may be that there really is no argument for mtDNA transfer that is not eugenicist; the reason to pursue it I outlined above rests on some sense of improving the stock of humanity, by doing what we can to eliminate an inherited illness. But if that is all there is to eugenics, there seems little to worry about. Reducing morbidity for future generations is a laudable thing; the puzzle then would be why eugenics is treated with suspicion at all.

Of course, the disastrous history of the mid-twentieth century provides us with a candidate answer to that. Eugenics as a programme has become contaminated by close association with racism and racists, and with systematic attempts to eliminate whole classes of people in the name of 'racial hygiene'; a recent article in *New Scientist* talks about sterilisation and genocide as 'the dark reality of eugenics', as though its moral aspect is a given (Bowler 2016). We should be wary of anything that might open the doors to this kind of behaviour. However, it is hard to see how mtDNA transfer has any conceptual links to racial superiority, and neither does it lend itself to extermination programmes or forced sterilisation. In short, if this is eugenics, then it looks like we might be able to reclaim the phrase: it could easily provide an example of what Nick Agar has called a 'liberal eugenics' – a form of eugenics that is chosen by women based on their conception of the good life, facilitated but not required by the state, and with no aspersions cast on anyone else (Agar 2004: 5). If we are determined to keep hold of the idea that making this kind of decision indicates a eugenicist mindset and is blameable, blame would presumably also have to attach to a woman who chooses not to conceive at all in order to avoid suboptimal health in future generations. Since I take this to be an argumentative non-starter, it is hard to see how choosing to avoid suboptimal health in future generations *while nevertheless reproducing* is blameable. Or, working the other way, a critic of mtDNA transfer may be able to keep hold of the claim that it is eugenic, but only at the cost of relinquishing the claim that eugenics is always morally reprehensible and never praiseworthy. And if the latter claim is relinquished, it's hard to see what would motivate the criticism.

A related line of argument to the appeal to eugenics comes from a version of the 'expressivist objection'. The objection is, in its purest form, a response to the possibility of selective abortion in the light of genetic tests, and relies on the idea that to choose to abort expresses undesirable attitudes not about a given trait, but about the person, and the kind of person, carrying it (Parens and Asch 1999). The objection can be modified to apply beyond abortion debates. Granted that a person can only exist thanks to the genome they have, a preference for a different genetic inheritance may express a preference for different people from the ones who do or may exist. This objection has its critics; but if there's anything to it, would it speak to mtDNA transfer? I am not convinced that it would. This is for two reasons, once conceptual and one empirical. The conceptual reason is that – as I indicated above – though some genetic alterations may be identity-affecting, it is not at all certain that a change in a person's mitochondrial inheritance is. I do not have time here to explore the metaphysics of identity, but I would hazard a guess that it is not, because of the relatively few genes that mitochondria carry. This being the case, altering someone's mtDNA is going to change the life of the person, not the identity person living it. For sure, we are wishing away mitochondrial illness; but we are *not* wishing away the person who has it at the same time.

And this speaks to the empirical reason why I think that the expressivist objection does not undermine the arguments for mtDNA transfer: it has significant support from people who carry faulty mitochondria, who would prefer that they or their offspring – extant or merely possible – didn't. The HFEA commissioned a report into mtDNA transfer in 2012, which involved focus groups drawn from patients; they supported the techniques (HFEA 2013). It is possible that such people are speaking in bad faith, or have internalised a dismissive attitude towards themselves, or something like that – but this would require independent evidence. Without that, the expressivist objection, at least in respect of mitochondrial illness, seems to fail.

In fine, the concerns about the freedom to manipulate a person's genetic inheritance being a preamble to a eugenicist's charter are either toothless, or never get going at all.

Neither should we forget that the law in the UK is what it is, and nothing more. It permits one specific kind of genetic engineering, in specific cir-cumstances, and permits that it be carried out only by a very few people.

It is true that any number of people have made the case for other kinds of genetic manipulation being permissible or even morally required (Harris 2007: ch. 3). But these are not allowed by any current legal instrument. If technologies such as CRISPR prove to be an effective and safe way to manipulate the genome, we may see moves in the appreciably near future to make use of it to fix other inherited illnesses. These proposals must be taken on their merits; it is probably fair to speculate that some would be acceptable, and others not. (An editorial in *Nature* in 2015 made a plea against making germline modifications to the human genome, largely because of the unknown risks – but, notably, made an exception for mtDNA transfer; Lanphier et al. 2015: 410–11.) None of this commits us to the view that liberalising the law in respect of *A* necessarily means that it will be liberalised in respect of *B*; and if it does happen to be liberalised in respect of *B*, it does not mean that all regulatory bets are off.

Freedom from illness

A little earlier, I made a passing comment that one may expect and require a reasonable polity to care about protecting the interests of its citizens and future citizens. This can be conceptualised as a guarantee of a certain kind of freedom – a freedom from exposure to risk. In this section, I want to explore that notion of freedom a little further in respect of mtDNA transfer.

We could treat mtDNA transfer as a kind of scientific freedom in the sense that it is a scientific technique that promises a kind of freedom to those on whose behalf it is performed. To be relieved of a burden is, at least colloquially, to be freed from it; it is perfectly natural to appeal to the language of freedom when talking about ending illness. For example, we might say that someone has been free from cancer for a certain time; conversely, we talk about the burden of illness. To add an adjective to this can be understood as describing the origin of this freedom; 'scientific freedom' is freedom that comes about by scientific means – and so also the removal of a burden by scientific means. There is no reason to think that mitochondrial illnesses would not fit into this linguistic pattern. Hence, if we think that one of the things that states should do is to maximise individuals' freedom (compatible with maintaining comparable levels of freedom for all), then it looks like we ought to applaud legal moves to remove restrictions from the use of mtDNA transfer. Freeing medicine to make use of the technique is a part of freeing individuals from the burden of mitochondrial disease.

Neither is it simply the individuals carrying faulty mitochondria who would be freed by this scientific endeavour. Illness and disability impose burdens on others – often close family members. The burdens may be more onerous in some cases than in others; and in most developed countries the state will have some role in providing medical and social assistance. Nevertheless, even the mundanity of extra doctors' appointments, filling in applications for assistance, and so on, will represent a drain on financial, psychological and social resources. The possibility of using mtDNA transfer to minimise the risk of mitochondrial illness therefore offers a way to free many people beyond the future child of any number of burdens.

Yet, on the other hand, there is also the possibility that this form of 'scientific freedom' may be less straightforward than it appears at first. Notably, undergoing the procedure is not 'cost free': it does require IVF, which is burdensome in its own right. Now, though a burden is something from which an agent can be freed, it does not follow from this that every burden is a restriction of freedom. One can willingly take on a burden without diminishing one's positive freedom. For example: choosing to pursue the dream of an Olympic medal brings with it a requirement to undergo a certain kind of training, but one is no less free for it. Though it's burdensome, the training is part and parcel of one's free choice. Likewise, a woman who felt it important not to pass faulty mitochondria to her children, but who didn't want to avoid sex altogether, would be no less free for choosing a combination of contraception and IVF (and she may, in passing, free subsequent generations from mitochondrial illness).

All this assumes, though, that a woman's desire for IVF is authentically hers. This is not a given. In examining a person's choices, it is important to keep in mind the social context in which those choices are made. I have already noted how the possibility of mitochondrial transfer adds a third option to a decision about whether to proceed with reproduction; and all else being equal, a free choice may be made. However, while considering a woman's choosing whether or not to reproduce, Chadwick observes that '[e]ither way, there are problems with blame and responsibility for the woman' (2009: 13). This phrase is telling. For most women, pregnancy by natural means is the default option, and goes completely unremarked; if it is the sort of thing in the context of which the word 'blame' starts to appear, this is no longer the case. Rather, the expectation would seem to be that a woman carrying faulty mitochondria ought either to remain celibate, to terminate any pregnancy that arises from a contraceptive failure, or to use mtDNA transfer. Natural reproduction for such women is not the default option; it's the exception, and a culpable one to boot. This being the case, our appeal to liberal eugenics might not be quite as straightforward as we may hope, for the supposedly free choice to make use of mtDNA transfer may be subject to social pressures that we do not often notice.

This worry cannot be dismissed as handwringing easily. In her *Genetic Dilemmas*, Dena Davis suggests that the availability of screening and termination for Down's syndrome shifts the range of courses of action seen as acceptable such that testing (and presumably termination: if you're going to have the child anyway, why screen?) become the default option, deviations from which are scrutable. Thus, she writes:

> In my own mind I can discern a subtle shift in the way in which I view people with certain anomalies. Twenty years ago, seeing a woman in the supermarket

with a child who has Down syndrome, my immediate reactions were sympathy
and a sense that that woman could be me. Now when I see such a mother
and child, especially if the mother is older, I am more likely to wonder why
she didn't get tested. (2001: 18)

Subtly, the norm has shifted; and if that shift can happen for this congenital
condition, it would be strange to insist that something similar cannot happen
in respect of mitochondrial illness and the choices that it brings. A woman's
reproductive decisions – or non-decisions, in the case of unplanned pregnancies – become matters of public comment. And this matters just because
the terms of a woman's decision about whether or not to use mtDNA
transfer is, howsoever subtly, altered.

To this, it may be pointed out that many or most pregnancies, especially
but not exclusively those in which there is a risk of some congenital health
problem, become matters of public comment; this is a bare feature of the
culture in which we live. This is may very well be true – in which case, so
much the worse for the idea that a woman's decision about her pregnancy
is free from sociopolitical influence. Since most women who carry defective
mtDNA know that they do, they may be under pressure similar to that
which Davis describes. No reproductive decision is free from influence or
scrutiny. This might mean that no reproductive decision is free, or that we
have to rethink what it is to be 'free' in respect of reproduction. A decision
about mtDNA transfer, though, probably isn't *sui generis*.

Whatever freedoms from illness mtDNA transfer promises, it may not
guarantee freedom in every possible sense. In some cases, it might even
generate social coercion, howsoever subtle it may be. I do not mean this
to imply any particular normative position. Sometimes moral pressure on
an agent is perfectly acceptable, if that agent would not otherwise act as
morality requires and the pressure is proportionate, even if that pressure
comes at the expense of the agent's freedom in part or in whole. It is not
difficult to come up with thought experiments in which this is the case. I
leave for others to decide whether the value of a future generation's freedom
from mitochondrial illness is worth the partial sacrifice of a potential mother's
moral freedom. All that matters for my purposes is to point out that 'scientific
freedom' in the sense of lifting one kind of burden may bring another kind
of burden with it.

Freedom to legislate

Permitting mtDNA transfer gave scientists and medics a freedom to carry
out the technique that they had not had before; it gave parents the freedom
to choose that technique; it gave future children a freedom from illness;
but it might have created a subtle pressure on some mothers-to-be. This
pressure should be taken seriously; but being careful about its exertion does
not mean that mtDNA transfer cannot be offered without it.

There is one final aspect of freedom in this context that is worth mentioning. The *Daily Telegraph* reported in 2015 that a group of fifty MEPs had written an open letter to David Cameron, decrying any liberalisation of the law on mtDNA transfer. The letter expressed a 'profound concern at the intention of the UK to permit the modification of the human genome', and continued to say that

> [y]our proposals violate the fundamental standards of human dignity and integrity of the person. Modification of the genome is unethical and cannot be permitted.
>
> These proposals put the UK out in front of a race to the bottom so far as standards of human dignity are concerned. (Knapton 2015)[2]

Were this the case, one might wonder whether the UK Parliament had overreached itself – if, that is, there are some things that are beyond the scope of what any national legislature can allow. Certainly the UN, in the Universal Declaration on the Human Genome and Human Rights (art. 1) has suggested that the human genome is part of the common heritage of humanity. If something is part of the common heritage of humanity, then it would seem to lie outside of national legislatures' demesne.

But, of course, dignity claims are notoriously slippery: what, exactly, does it mean to violate human dignity? Without a coherent account of human dignity, we cannot begin to assess that claim – let alone be expected to accept it. Likewise, worries about personal integrity are hard to understand for very similar reasons. Finally, it seems perfectly reasonable to deny that the *human* genome would be altered by mtDNA transfer, because no genes from another organism nor any synthetic genes are implanted. Besides, it is possible to insist that mtDNA transfer no more alters the human genome than does nature anyway: hundreds of pathogenic mtDNA mutations have been identified, and mtDNA in humans has a very high rate of natural mutation compared to nuclear DNA (Tuppen 2010: 115; Lane 2015: 226–7). A 'pristine' human genome would be a rather Platonic thing, unaffected by the alterations that make each of us what we are; but if we are going to allow that such a thing exists, we might easily hypothesise that mtDNA transfer restores the human genome to something closer to that ideal.

On a much more rough-and-ready level, though, we might think it enough to point out that mtDNA transfer promises to relieve great suffering, and that that is enough to mean that legislatures that seek to permit it should feel free to do so.

Notes

1 As I write this, it is reported that the first baby to be born after mtDNA transfer has been delivered in Mexico, which was apparently chosen because 'there are no rules'. See Hamzelou (2016: 8–9).

2 The letter is widely reported, but I have not been able to track down the original.

References

Agar, N. (2004), *Liberal Eugenics*, Malden, MA: Blackwell.

Bowler, P. (2016), 'From genetics to eugenics', *New Scientist*, 231.3088: 42.

Brassington, I. (2013), *Bioscience and the Good Life*, London: Bloomsbury.

Chadwick, R. (2009), 'Gender and the human genome', *Mens Sana Monographs*, 7.1: 13.

Darwin, L. (1929), *What Is Eugenics?*, London: Watts.

Davis, D. (2001), *Genetic Dilemmas*, New York: Routledge.

Ellison, J. (2015), HC Deb. 3 February 2015, vol. 592, col. 163.

Flello, R. (2014), HC Deb. 12 March 2014, vol. 577, col. 166WH.

Harris, J. (2007), *Enhancing Evolution*, Princeton, NJ: Princeton University Press.

Hamzelou, J. (2016), '"3-parent baby" success', *New Scientist*, 232.3093: 8–9.

HFEA (2013), 'Medical frontiers: debating mitochondria replacement, Annex VI: Patient Focus Group', London: OPM, www.hfea.gov.uk/9359.html (last accessed 10 May 2017).

Human Fertilisation and Embryology (Mitochondrial Donation) Regulations (2015), www.legislation.gov.uk/uksi/2015/572/pdfs/uksi_20150572_en.pdf (last accessed 10 May 2017).

Knapton, S. (2015), 'Three parent babies: Britain has breached EU law, MEPs warn', *Daily Telegraph*, 21 February.

Lane, N. (2015), *The Vital Question*, London: Profile.

Lanphier, E., Urnov, F., and Haeckner, S. E. et al. (2015), 'Don't edit the human germ line', *Nature*, 519: 410–11.

Liao, S. M. (2017), 'Do mitochondrial replacement techniques affect qualitative or numerical identity?', *Bioethics*, 31.1: 20–6.

Mill, J. S. (1985 [1859]), *On Liberty*, London: Penguin.

Newman, S. (2013), 'The British embryo authority and the chamber of eugenics', *Huffington Post*, 11 May, www.huffingtonpost.com/stuart-a-newman/mitochondrial-replacement-ethics_b_2837818.html (last accessed 10 May 2017).

Parens, E., and Asch, A. (1999), 'The disability rights critique of prenatal genetic testing: reflections and recommendations', *Hastings Center Report*, 29.5: S2.

Rulli, T. (2017), 'The mitochondrial replacement "therapy" myth', *Bioethics*, doi: 10.1111/bioe12332.

Tuppen, A., Blakely, E. L., and Turnbull, D. M. et al. (2010), 'Mitochondrial DNA mutations and human disease', *Biochimica et Biophysica Acta (BBA) – Bioenergetics*, 1797.2: 116–17.

Wrigley, A., Wilkinson, S., and Appleby, J. B. et al. (2015), 'Mitochondrial replacement: ethics and identity', *Bioethics*, 29.9: 631–8.

6

Scientific freedom and responsibility in a biosecurity context

Catherine Rhodes

Scientific freedoms are exercised within the context of certain responsibilities, which in some cases justify constraints on those freedoms. (Constraints that may be internally established within scientific communities and/or externally enacted.) Biosecurity dimensions of work involving pathogens are one such case and raise complex challenges for science and policy. The central issues and debates are illustrated well in the development of responses to publication of ('gain of function') research involving highly pathogenic avian influenza, by a number of actors, including scientists, journal editors, scientific academies, and national and international policy groups.

The core tension that can arise between working to protect health by promoting work on pathogens (to support surveillance and response efforts) and working to protect health by setting limits to work on pathogens that poses risks to health through accidental or deliberate releases, is reflected in and has been responded to by international (and some national) policy processes. These responses have placed increasing emphasis on the responsibilities of scientists. Framed within recognition of the reciprocal responsibilities of scientists and policymakers, further joint work is needed to manage this tension and develop appropriate and effective international responses.

Scientific freedoms and (scientific) responsibilities

Scientific freedom is subject to internal and external constraints, some of which relate to responsibilities that are widely recognised by the scientific community. This chapter focuses on international dimensions of science–policy interactions, and the conception of the scientific community used in this chapter reflects this, and fits the definition provided by Henk Verhoog (1981: 583): 'the community of scientific workers wherever they are in the world, sharing the same general conception of nature and the same basic methodological norms'. The responsibilities discussed in this chapter may be assigned to individual scientists, to groups working in particular areas, and/or to the scientific community as a whole.

Broadening of scientific responsibilities

Scientific responsibility has both internal and external dimensions. The internal dimensions can be thought of as the responsibilities that scientists have towards the scientific community and to upholding good scientific practice, and relate to traditional elements of 'responsible conduct of research'. Frankel and Carlson (2011), for example, list the following areas covering the main elements of internal scientific responsibilities:

- data acquisition, management, sharing and ownership;
- conflict of interest and commitment;
- human subjects;
- animal welfare;
- research misconduct;
- publication practices and responsible authorship;
- mentor/trainee responsibilities;
- peer review; and
- collaborative science.

The more traditional conception of scientific responsibility has broadened to include external dimensions – because of the important relationship that science has with and the implications it has for society – and in line with recognition of the increasingly global nature of science (National Academies 2009; Rhodes and Sulston 2010; InterAcademy Council 2012). These more outward-facing responsibilities include consideration of research outcomes and alignment of research goals and plans with societal concerns. The German Ethics Council (2014: 57) makes similar arguments, for example stating that:

> [T]he sciences are increasingly understood as being not merely self-contained processes that take place within a scientific community, but rather as being an integral component of general societal interrelationships . . . Accordingly, the ethical appraisal of science must concern itself not only with the practical consequences of knowledge, but also with the effects of the research process and its findings on society.

Biosecurity concerns relate to the dual-use nature of some research, and this appears primarily to relate to the external dimension of scientific responsibility – relating to avoidance of harm to society through the deliberate misuse of research 'to pose a threat to public health and safety, agricultural crops and other plants, animals, the environment or materiel' (NSABB n.d.). Making clear the connection of biosecurity concerns to internal norms of responsible conduct is also important, and modifications to e.g. laboratory practices may be necessary (WHO 2013).

Some of the publications providing guidance on responsible conduct of research have incorporated certain aspects of biosecurity into internal responsibilities. For example, the InterAcademy Council's 2012 Report

– *Responsible Conduct in the Global Research Enterprise*, includes 'misuse of biological agents' within its list of irresponsible research practices (5).

In recent years some countries, for example the Netherlands and Germany have produced more specific guidance on scientific responsibility and biosecurity (Royal Netherlands Academy of Arts and Sciences 2013; German Ethics Council 2014). However, such activities are not widespread, and it is unclear whether the scientific community has a high level of awareness of such guidance and what it might mean in practice, nor how broadly such responsibilities are accepted as an extension of internal norms in the scientific community.

Appropriate handling of biosecurity will also help to ensure that public trust in science is not jeopardised – which is an important general motivation for upholding standards of responsible scientific conduct. This is clearly stated in the World Health Organization's *Laboratory Biosecurity Guidance* (2006: 2, 25)

> The general public expects laboratory personnel to act responsibly and not to expose the community to biorisks, to follow safe working practices (biosafety) associated with practices that will help keep their work and materials safe and secure (biosecurity).
>
> Effective laboratory biosecurity is a societal value that underwrites public confidence in biological science.

This was also a motivation for the Biotechnology and Biological Sciences Research Council (BBSRC), Medical Research Council (MRC) and Wellcome Trust's *Position Statement on Dual Use Research of Concern and Research Misuse* (2015: para. 22).

Biosecurity responsibilities relate not only to ensuring that individuals do not follow intentionally harmful/prohibited lines of research, but also to consideration of how research might be used by others, with a requirement in certain cases to modify research plans or communication activities, and even to forgo certain lines of research altogether, when there is a significant risk of misuse. It is generally agreed that this will only apply to a very small subset of research (Royal Society 2012: 57; DFG and Leopoldina 2014: 10; BBSRC et al. 2015: para. 9; NSABB 2016: 1).

Guidance documents and statements produced by scientific academies and funding bodies relating to scientific responsibility place strong emphasis on the fundamental importance of openly communicating research – to contribute to scientific progress, to enable replication and verification of findings, etc. Connected to this, the ability to openly communicate is viewed as a core scientific freedom. The importance of this aspect of scientific freedom and responsibility is particularly significant in terms of the discussion in this chapter. Among many examples of such statements:

> Open communication and deliberation sit at the heart of scientific practice.
> (Royal Society 2012: 13)

[P]ublication in a peer-reviewed journal is the most important way of disseminating a complete set of research results. (National Academies 2009:
29)

Researchers who fail to meet these expectations place their reputations at
risk. (InterAcademy Council 2012: 16)

The free exchange of information and especially the publication of results are
important factors for scientific knowledge and scientific progress. (DFG
and Leopoldina 2014: 13)

Some more recent documents have qualified this expectation slightly
(Royal Society 2012; DFG and Leopoldina 2014), but it is not clear whether
this is a generally accepted modification to the core scientific norm of open
communication – and it is not clear how this should be balanced with
apparently contradictory statements. The following statements, for example,
appear in the Royal Society's *Science as an Open Enterprise* report (2012:
9, 57):

Qualified openness
Opening up scientific data is not an unqualified good. There are legitimate
boundaries of openness which must be maintained in order to protect
commercial value, privacy, safety and security. Careful scrutiny of the
boundaries of openness is important where research could in principle be
misused to threaten security, public safety or health.

A joint report by the Royal Society, the InterAcademy Panel and the International
Council of Science in 2006 concluded that 'restricting the free flow of
information about new scientific and technological advances is highly
unlikely to prevent potential misuse and might even encourage misuse'.

Reciprocal responsibilities of science and society

In general, the relationship between science and society can be described
as reciprocal, and there are responsibilities for both sides that are associated
with this. Scientists have responsibilities towards society inter alia because
science contributes to a range of social goods, because it has significant
social and economic impacts, and because it is ultimately funded by society,
whether through taxation or consumer spending (iSEI 2010: 3; InterAcademy
Council 2012: v). These responsibilities include: addressing the concerns,
values and interests of society; consideration of the implications of their
work for society; providing policy advice in areas in which they are qualified
to do so; and accepting some limitations on scientific freedom or conditions
on practice to align with social goals and values.

Society as a collective recipient of benefits from scientific research, has
responsibilities towards it, for example in facilitating scientific progress
through e.g. funding, provision of education and training, and enabling
dialogue and scrutiny (e.g. through broad scientific literacy), and by generally protecting scientific freedoms where these don't substantially conflict
with other protected values. In other words, 'society needs to provide

just and effective conditions for the increase of scientific knowledge' (iSEI 2010: 3).

Policymakers are important intermediaries in the reciprocal relationship between science and society. They need to respond to and where necessary achieve balance between the interests and values of both groups – utilising a range of policy tools (e.g. regulation, licensing, guidance, funding). Scientists often have specialist expertise that can inform policymaking, and should seek to provide appropriate input (InterAcademy Council 2012: 28). Policymakers have a responsibility to ensure that evidence/advice is sought and to consider the impacts of policy for scientific practice. This will, however, be in the context of a variety of constraints and demands on policymaking, and limitations on policymakers' freedom need to be acknowledged, in order to establish realistic expectations of science–policy relationships. This point is picked up on in the World Health Organization's guidance *Responsible Life Sciences Research*: 'One the one hand, government policy should aim to promote the advancement of science. Scientific progress usually has important societal (including economic) benefits; and promoting the good of society is a primary responsibility of government . . . At the same time, safety, security and economic development are significant responsibilities of governments' (WHO 2010: 28).

Tension between freedoms and responsibilities

There are many different situations in which freedoms and responsibilities may be in tension and need to be carefully balanced. In the biosecurity context a core tension arises because some research which aims to protect health can also create health risks, particularly research involving work on dangerous pathogens. Such research serves to protect health (e.g. by supporting preparedness and surveillance efforts, the development of treatments, and other public health responses to outbreaks). Through unintentional or deliberate actions such work can also threaten health e.g. through accidental release of a pathogen, or through deliberate misuse e.g. to create weapons. It is the latter – deliberate misuse (of materials and data and knowledge emerging from research) that is a central concern in biosecurity, and the main focus of discussion in this chapter.

Scientific advances tend to contribute to both aspects of this tension, because they can serve both to facilitate misuse (e.g. by enabling creation of more desirable warfare agents, such as pathogens that are more virulent, environmentally stable, or for which there are no prophylactic measures available), and to enhance defence measures (e.g. through new detection tools), and assist identification and monitoring of, and response to outbreaks.

The biosecurity context

The term biosecurity has various meanings in different contexts.[1] Its use in this chapter covers prevention of the deliberate misuse of biological (and related) sciences to cause harm. While most scientific work can have dual-use implications,[2] certain types of research are viewed as presenting particularly significant biosecurity risks, and have been subject to particular policy attention. The types of research of concern are frequently linked in policy to research on particular agents (including those associated with past biological weapons programmes) and to particular types of experiment. This is the approach, for example, used in the US government's *Policy for the Oversight of Dual Use Research of Concern* (US Government 2012). A widely used example of types of 'experiments of concern' is found in the 2004 report *Biotechnology Research in an Age of Terrorism* (National Research Council of the National Academies 2004):

1. Would demonstrate how to render a vaccine ineffective.
2. Would confer resistance to therapeutically useful antibiotics or antiviral agents.
3. Would enhance the virulence of a pathogen or render a non-pathogen virulent.
4. Would increase the transmissibility of a pathogen.
5. Would alter the host range of a pathogen.
6. Would enable evasion of diagnostic/detection modalities.
7. Would enable the weaponisation of a biological agent or toxin.

Box 2 in *Position Statement on Dual Use Research of Concern and Research Misuse* (BBSRC et al. 2015: 2) suggests the addition of:

[E]xperiments that would: generate or reconstitute an eradicated or extinct agent or toxin.

[T]he development of new technologies or tools with genetic applications – such as in the areas of bio-processing or bio-fermentation scale-up – which could, for example, make it easier to synthesise or produce harmful agents.

[P]rojects that carry very little potential for misuse, but where the risk would be greatly increased by emerging data or methodologies from other disciplines, for example studies on a toxin that cannot be introduced easily into humans, but which might be deliverable by advances in materials science or aerosol physics.

'Gain of function' research

The way and the extent to which biosecurity considerations might limit research on pathogens has gained a high profile among some scientific, security and policymaking groups over the past few years, particularly because of debate around the publication of highly pathogenic avian influenza (H5N1) 'gain of function'[3] experiments.

Two groups – led by Ron Fouchier and Yoshihiro Kawaoka – had been conducting research to identify mutations to the H5N1 influenza virus that could make it transmissible between humans. The work aimed to assist in the identification of strains likely to cause pandemics and in the development of vaccines and treatments. The groups submitted manuscripts to *Nature* and *Science* in 2011. As the research had received US federal funding (through the National Institutes for Health), the National Science Advisory Board for Biosecurity (NSABB) reviewed the manuscripts before publication. The NSABB is an advisory committee for the US government, inter alia addressing 'policies governing publication, public communication, and dissemination of dual use research methodologies and results' (US Department of Health and Human Services 2016: 1).

In giving its recommendations on the original transcripts, the NSABB noted that the publications 'described the generation of mutations in H5N1 that enable the airborne transmission of the virus between ferrets'; 'recognized the importance of the research in advancing knowledge of influenza transmission and supporting public health efforts'; but noted that 'specific findings would enable others to synthesize and express a H5N1 strain with mammal-to-mammal transmissibility' and that they had 'significant concerns that information in the manuscripts could be misused to endanger public health and national security'. They therefore recommended 'the information in these manuscripts be published in a redacted form with the omission of certain details that could enable the direct misuse of the research by those with malevolent intent' (2012: 1).

The NSABB reviewed revised versions of the manuscripts in March 2012, along with 'new non-public epidemiological information, and security information . . . presented in a classified briefing' (2012: 1). While it still viewed the manuscripts as presenting 'dual use research of concern', a majority concluded that:

> As currently written, the revised manuscripts do not appear to provide information that would enable the near-term misuse of the research . . . The mutations described in the manuscripts do not appear to result in H5N1 viruses that are both highly pathogenic and transmissible between ferrets through the air . . . The revised manuscripts provided a greater appreciation of the direct applicability of the information to ongoing and future influenza surveillance efforts. (NSABB 2012: 2–3)

It recommended that 'the revised Kawaoka manuscript be communicated in full' and that – with some further revisions, the Fouchier manuscript could also be published (NSABB 2012: 5). The manuscripts were published in *Nature* and *Science* later in 2012.

Alongside this process, scientists working on these types of H5N1 influenza transmission studies held a voluntary moratorium, to give space for further discussion and debate. This lasted around one year; with the researchers announcing in early 2013 that its aims had largely been met and so research would continue (Fouchier et al. 2013).

There have been additional policy responses in the US, including advice issued in 2013 on enhanced biosafety requirements for research on highly pathogenic H5N1 viruses (US Department of Health and Human Services, National Institutes of Health 2013) and – announced in October 2014 – a research funding pause 'on selected gain of function research involving influenza, MERS and SARS viruses' during a deliberative process towards recommendations that would 'inform the development and adoption of a new US government policy governing the funding and conduct of gain-of-function research' (US Government 2014). This included public consultation, a risk–benefit analysis, and an ethical analysis of such research, contributing to recommendations from the NSABB. The recommendations were published in May 2016, and they advise subjecting 'a small subset of GOF research – GOF research of concern (GOFROC)' to additional review and oversight (NSABB 2016: 1).

> Research proposals involving GOF research of concern entail significant potential risks and should receive an additional, multidisciplinary review, prior to determining whether they are acceptable for funding. If funded such projects should be subject to ongoing oversight at the federal and institutional levels. (NSABB 2016: 2)

The NSABB also outlined particular attributes of 'gain of function research of concern' and a set of principles to guide funding decisions (2016: 43–4).

There are a few other examples of additional guidance being provided in response to the influenza transmission studies, including the Royal Netherlands Academy of Arts and Sciences report *Improving Biosecurity: Assessment of Dual Use Research* (2013), and the German Ethics Council's opinion *Biosecurity – Freedom and Responsibility of Research* (2014). The World Health Organization (WHO) has hosted technical discussions on the Fouchier and Kawaoka studies (WHO 2012), and on dual-use research of concern more broadly (WHO 2013). A coordinated international approach to the funding, conduct, oversight and communication of such experiments – although recognised to be necessary given the global nature of the scientific enterprise and the public health threat – has yet to be developed.

International governance

There is a long-standing international recognition of the need to protect human, animal and plant life and health from transboundary disease threats. Within these efforts it is recognised that international scientific collaboration is essential for appropriate and effective monitoring, surveillance and response efforts (OIE 2015; WHO 2010). Biosecurity issues sit at the intersection of security and health and there are relevant international rules, organisations and other governance mechanisms in both domains, key components of which are outlined below.

The Biological Weapons Convention

The Biological Weapons Convention (BWC) is based around a core prohibition in Article I:

> Each State Party to this Convention undertakes never in any circumstance to develop, produce, stockpile or otherwise acquire or retain:

> (1) Microbial or other biological agents or toxins, whatever their origin or method of production, of types or in quantities that have no justification for prophylactic, protective or other peaceful purposes. (United Nations Biological Weapons Convention 1972)

Notably, the Convention does not seek to restrict use of biological agents and toxins for peaceful purposes; it explicitly encourages participation and international cooperation for such purposes, particularly including the prevention of disease:

> (1) The States Parties to this Convention undertake to facilitate, and have the right to participate in, the fullest possible exchange of equipment, materials and scientific and technological information for the use of bacteriological (biological) agents and toxins for peaceful purposes. Parties to the Convention in a position to do so shall also cooperate in contributing individually or together with other States or international organisations to further development and application of scientific discoveries in the field of bacteriological (biological) for the prevention of disease, or for other peaceful purposes. (Article X)

States parties to the BWC track scientific and technological developments relevant to the Convention, including those 'that have potential for uses contrary to the provisions of the Convention' and those 'that have potential benefits for the Convention, including those of special relevance to disease surveillance, diagnosis and mitigation' (United Nations 2012: 23).

Other relevant governance efforts that fall within the security domain include work under United Nations Security Council Resolution 1540, which 'Obliges States, inter alia, to refrain from supporting by any means non-State actors from developing, acquiring, manufacturing, possessing, transporting, transferring or using nuclear, chemical or biological weapons and their delivery systems' (United Nations 1540 Committee, n.d.) and informal export control arrangements of the Australia Group (Government of Australia 2007).

Health and disease control efforts

There are international rules and systems for the protection of human, animal and plant life and health. Those which seek to prevent the introduction and spread of infectious disease are particularly relevant, and include:

- The international organisations tasked with protection of human, animal and plant health – the World Health Organization, the World Animal Health Organization (OIE), and the Food and Agriculture Organization (FAO).
- The main rules seeking to control the movement of pests and diseases through international travel and trade routes (which also establish requirements e.g. for reporting and managing outbreaks of international concern) – the International Health Regulations, the Terrestrial and Aquatic Animal Health Codes and Manuals, and the International Plant Protection Convention.
- Specific guidance on laboratory biosafety and biosecurity, to minimise the risks of accidental or deliberate release of pathogens from laboratories and the risks to health of laboratory workers.
- Guidance on the transport of infectious substances (generally found within dangerous goods regulations issued separately for each transport mode, but with key parts summarised in WHO's *Guidance on Regulations for the Safe Transport of Infectious Substances*).
- A range of surveillance and response systems – including more general systems such as WHO's Global Outbreak Alert and Response Network and OIE's World Animal Health Information System; more specific systems such as WHO's Global Influenza Surveillance and Response System.
- Expert and laboratory networks including WHO's and OIE's collaborating centres and reference laboratories, and joint systems such as OFFLU, a collaboration of the OIE and FAO networks on animal influenzas.

The disease surveillance and response systems and collaborative networks, in particular, rely on timely and effective sharing of materials, data and research.

While these organisations primarily focus on natural disease threats, they recognise that their systems will play an essential role in the identification of and response to any deliberately caused disease outbreaks. The OIE, for example, addressed these issues within its Biological Threat Reduction Strategy: 'The same disease surveillance and intelligence systems that are in place to detect day-to-day occurrences of natural outbreaks in animals, within countries and at national borders, will also detect deliberate and accidental releases' (OIE 2015: 3).

Handling the tension

The tension between protecting health through access to pathogens and open communication of research results and protecting health by restricting access to pathogens and associated data and knowledge with misuse potential is being addressed within these organisations and by states parties to the BWC. In their responses to the misuse potential of work involving pathogens, in recent years they have placed increasing emphasis on the responsibilities of scientists. This includes recommendations about building awareness among scientific communities of their legal and moral obligations, development of codes of conduct and ethics education and training, and the creation of a culture of responsible science, in which scientists are able to make informed decisions about the biosecurity aspects of their research (which may in some cases constrain scientific freedoms e.g. by limiting or delaying communication of research, or not following certain lines of enquiry).

For example, the Seventh Review Conference of the BWC noted 'the value of national implementation measures' which 'implement voluntary management standards on biosafety and biosecurity'; 'promote the development of training and educational programs' and 'encourage the promotion of a culture of responsibility amongst relevant national professionals and the voluntary development, adoption and promulgation of codes of conduct' (United Nations 2012: art. IV, para. 13, secs. a, d and e). And its series of intersessional meetings from 2012 to 2015 included in the standing agenda item on review of scientific and technological developments, the topics of 'voluntary codes of conduct and other measures to encourage responsible conduct by scientists, academia and industry' and 'education and awareness raising about risks and benefits of life sciences and biotechnology' (2012: para. 22, secs. d and e).

The WHO's *Laboratory Biosecurity Guidance* included recommendations on development of codes of conduct for researchers and other laboratory staff, which ought to include: 'evaluation of the purpose of the work, consideration for its impact the publication of research results, and [enumeration of] considerations and conditions for or against the publication of results that may have dual-use implications' (2006: 21); and that there is a need for promotion of 'a culture of awareness, shared sense of responsibility, ethics and respect of codes of conduct within the international life sciences community' (2006: 30). The WHO further recommended that 'comprehensive bioethical reviews should be carried out and documented before final decisions are reached on the publication of data, balancing pros and cons of their dissemination' (2006: 21). (Interestingly it then went on to suggest that one particular area in which further examination of 'bioethical considerations, international review and control of this research' was needed was in studies 'with the highly pathogenic H5N1 to investigate virus transmissibility' (2006: 21).) (The WHO also produced a guidance document *Responsible Life Sciences Research for Global Health Security* in 2010, which provides a public health perspective on such issues.)

The OIE's Biological Threat Reduction Strategy includes aims to: 'Maintain the OIE laboratory twinning programme to improve compliance with OIE Intergovernmental Standards, including for biosafety and biosecurity, to create a culture of responsible science and good laboratory practice', including through incorporation in graduate training; and

> Advocate that fostering altruistic scientific networks at the national, regional and global level is a means of sustaining expertise, and preventing scientists from contributing to bioweapons development, by encouraging a culture of responsible and transparent science. (OIE 2015: 1)

Conclusion: shared and reciprocal responsibilities in the biosecurity context

These recent international efforts to respond to the tension between scientific freedoms and responsibilities in the biosecurity context are positive steps towards raising scientific and policy awareness of the issues. However, recent cases such as the H5N1 mammalian transmissibility studies, have highlighted significant policy gaps at a national and international level, and indicate that awareness raising work and science–policy engagement on the development and implementation of biosecurity measures, need substantial further work.

While some of these efforts will necessarily involve consideration and action by individual scientists, with strong institutional support, it is important to recognise that addressing biosecurity issues relating to research, is not a responsibility of the scientific community alone. It can be framed within the reciprocal relationship of science and society outlined above, as a shared responsibility – particularly, within the context of this chapter – of the scientific community and international policymakers in relation to the development and implementation of biosecurity initiatives. Such initiatives are likely to entail some constraints being placed on scientific freedoms by policymakers on security grounds.

The science–society relationship in this area is mediated by other actors, who also have important roles to play in the development of appropriate policy, including research institutions, scientific academies, funders and publishers (United Nations 2013; German Ethics Council 2014: 52). Responsibilities shared by these groups include enabling scientists to make informed judgements on issues with which they may currently be unfamiliar, developing guidance to support decision-making processes at the institutional level, and contributing to effective engagement and deliberation among all groups about the appropriate balance between various values and interests, and robust and broadly acceptable ways of achieving that balance.

> Responsible decision-making is required by actors at all levels. Decision-makers will need to make judgements to resolve difficult cases of conflicting values. Scientific freedom, scientific progress, public health, safety and security are all important values and none should be given absolute priority over the others. (WHO 2010: 28)

Policymakers' freedom of action also faces various constraints, including lack of knowledge, expertise and evidence. They therefore need, and policy will benefit from, appropriate and effective input from experts, including better information about scientific and technological advances, the impact these can have on the operation of international rules, and the impact that such rules can have on scientific practice (Selgelid 2009). In the biosecurity area the expertise needed will include science, security, public health and ethics, among other areas; there are established methods and processes for drawing in expertise to international policy processes, but these need adequate and sustainable support (Rhodes 2014).

In turn, the scientific community needs to be better informed about the biosecurity implications of research (WHO 2010: 26), and about international governance and policy processes and the ways in which it can engage with them. This will require ongoing dialogue and improvement of existing mechanisms for scientific review and advice. Notably, improvements to the science and technology review process for the BWC were considered by states parties at the Eighth Review Conference in 2016, but little progress was made.

The expectation of responsibility for the consequences of research is relatively uncontroversial for intended applications and easily foreseeable outcomes, but becomes more difficult when assigning responsibility for use of research by others, in unintended ways. Even where scientists do not accept the argument that awareness of and response to biosecurity concerns is part of their responsibilities, their engagement in such processes is still very important, if only from a self-interested perspective, because otherwise policy will develop and be implemented without appropriate scientific input.

Notes

1 This is for example discussed in German Ethics Council (2014: 11–14).
2 In relation to research that has relevance for biosecurity, dual use refers to 'research conducted for legitimate purposes that generates knowledge, information, technologies, and/or products that can be utilised for both benevolent and harmful purposes' (NIH 2014).
3 'Gain of function' refers to a much broader set of research than that which raises biosecurity concerns – see NSABB (2016: 5, Box 1).

References

BBSRC (Biotechnology and Biological Sciences Research Council), MRC (Medical Research Council) and Wellcome Trust (2015), 'Position Statement on dual use research of concern and research misuse', https://wellcome.ac.uk/sites/default/files/wtp059491.pdf (last accessed 8 October 2016).

DFG [Deutsche Forschungsgemeinschaft] and Leopoldina (2014), *Scientific Freedom and Scientific Responsibility: Recommendations for Handling Security-relevant Research*, www.leopoldina.org/uploads/tx_leopublication/2014_06_DFG-Leopoldina_Scientific_Freedom_Responsibility_EN.pdf (last accessed 8 October 2016).

Fouchier, R., García-Sastre, A., and Kawaoka, Y. et al. (2013), 'Transmission studies resume for avian flu', *Science*, 339.6119: 520–1.

Frankel, M. S., and Carlson, R. (2011), 'Reshaping responsible conduct of research', *American Association for the Advancement of Science Professional Ethics Review*, 24.1, www.aaas.org/sites/default/files/migrate/uploads/Professional-Ethics-Report-Delta.pdf (last accessed 8 October 2016).

German Ethics Council (2014), *Biosecurity – Freedom and Responsibility of Research: Opinion*, Berlin: Deutscher Ethikrat, www.ethikrat.org/files/opinion-biosecurity.pdf (last accessed 8 October 2016).

Government of Australia (2007), Australia Group, www.australiagroup.net/en/index.html (last accessed 8 October 2016).

InterAcademy Council (2012), *Responsible Conduct in the Global Research Enterprise: A Policy Report*, Amsterdam, IAP, www.interacademies.net/file.aspx?id=19789 (last accessed 8 October 2016).

iSEI [Institute for Science, Ethics and Innovation] (2010), 'Who owns science? The Manchester Manifesto', www.isei.manchester.ac.uk/TheManchesterManifesto.pdf (last accessed 8 October 2016).

National Academies (2009), *On Being A Scientist: A Guide to Responsible Conduct in Research, 3rd Edition*, Washington, DC: National Academies Press, www.nap.edu/catalog/12192/on-being-a-scientist-a-guide-to-responsible-conduct-in (last accessed 8 October 2016).

National Institutes of Health (2014), Tools for the Identification, Assessment, Management and Responsible Communication of Dual Use Research of Concern: A Companion Guide to United States Government Policies for Oversight of Life Sciences Dual Use Research of Concern, www.phe.gov/s3/dualuse/Documents/durc-companion-guide.pdf (last accessed 8 October 2016).

National Research Council of the National Academies (2004), *Biotechnology Research in an Age of Terrorism*, Washington, DC: National Academies Press, www.nap.edu/catalog/10827/biotechnology-research-in-an-age-of-terrorism (last accessed 8 October 2016).

NSABB [National Science Advisory Board for Biosecurity] (2012), *Findings and Recommendations: March 29–30, 2012*, http://osp.od.nih.gov/sites/default/files/resources/03302012_NSABB_Recommendations_1.pdf (last accessed 8 October 2016).

NSABB [National Science Advisory Board for Biosecurity] (2016), *Recommendations for the Evaluation and Oversight of Proposed Gain of Function Research*, http://osp.od.nih.gov/sites/default/files/resources/NSABB_Final_Report_Recommendations_Evaluation_Oversight_Proposed_Gain_of_Function_Research.pdf (last accessed 8 October 2016).

NSABB [National Science Advisory Board for Biosecurity] (n.d.), 'National Science Advisory Board for Biosecurity: frequently asked questions', http://osp.od.nih.gov/office-biotechnology-activities/biosecurity/nsabb/faq (last accessed 8 October 2016).

OIE [World Animal Health Organization] (2015) *Biological Threat Reduction Strategy: Strengthening Global Biological Security*, Paris, www.oie.int/fileadmin/Home/docs/pdf/EN_FINAL_Biothreat_Reduction_Strategy_OCT2015.pdf (last accessed 8 October 2016).

Rhodes, C. (2014), 'BTWC: learning from alternative models of science and technology review', *Policy Paper 7* for the Biochemical Security 2030 Project, http://biochemsec2030.org/policy-outputs (last accessed 8 October 2016).

Rhodes, C., and Sulston, J. (2010), 'Scientific responsibility and development', *European Journal of Development Research*, 22.1: 3–9.

Royal Netherlands Academy of Arts and Sciences [KNAW] Biosecurity Committee (2013), *Advisory Report – Improving Biosecurity: Assessment of Dual-Use Research*, Amsterdam: KNAW, www.knaw.nl/en/news/publications/improving-biosecurity (last accessed 8 October 2016).

Royal Society (2012), *Science as an Open Enterprise: The Royal Society Science Policy Centre Report 02/12*, London, https://royalsociety.org/~/media/policy/projects/sape/2012-06-20-saoe.pdf (last accessed 8 October 2016).

Selgelid, M. (2009), 'Governance of dual-use research: an ethical dilemma', *WHO Bulletin*, 87.9: 720–3, www.who.int/bulletin/volumes/87/9/08–051383/en (last accessed 8 October 2016).

United Nations (2012), *Final Document of the Seventh Review Conference*, w w w . u n o g . c h / 8 0 2 5 6 E D D 0 0 6 B 8 9 5 4 / (h t t p A s s e t s) /3E2A1AA4CF86184BC1257D960032AA4E/$file/BWC_CONF.VII_07+(E).pdf (last accessed 8 October 2016).

United Nations (2013), BWC/MSP/2013/WP.10, 'Addressing modern threats in the Biological Weapons Convention: a food for thought paper, submitted by Australia, Canada, Chile, Colombia, Czech Republic, Finland, Ghana, Lithuania, the Netherlands, Nigeria, Republic of Korea and Sweden, https://documents-dds-ny.un.org/doc/UNDOC/GEN/G13/645/78/PDF/G1364578.pdf?OpenElement (last accessed 8 October 2016).

United Nations 1540 Committee (n.d.), 'Factsheet', www.un.org/en/sc/1540/1540-fact-sheet.shtml (last accessed 8 October 2016).

United Nations Convention on the Prohibition of the Development, Production and Stockpiling of Bacteriological (Biological) and Toxin Weapons and on their Destruction (1972), www.unog.ch/80256EDD006B8954/(httpAssets)/C4048678A93B6934C1257188004848D0/$file/BWC-text-English.pdf (last accessed 8 October 2016).

US Department of Health and Human Services (2016), *National Science Advisory Board for Biosecurity Charter*, http://osp.od.nih.gov/sites/default/files/NSABB_Charter_2016.pdf (last accessed 8 October 2016).

US Department of Health and Human Services, National Institutes of Health (2013), 'Recombinant DNA research: actions under the NIH guidelines for research involving recombinant DNA molecules', *Federal Register*, 78.35: 12074–82, http://osp.od.nih.gov/sites/default/files/resources/FR%202_21_2013_78_FR_12074.pdf (last accessed 8 October 2016).

US Government (2012), *Policy for the Oversight of Dual Use Research of Concern*, http://osp.od.nih.gov/sites/default/files/resources/United_States_Government_Policy_for_Oversight_of_DURC_FINAL_version_032812_1.pdf (last accessed 8 October 2016).

US Government (2014), Deliberative Process and Research Funding Pause on Selected Gain-of-function Research Involving Influenza, MERS and SARS Viruses, www.phe.gov/s3/dualuse/documents/gain-of-function.pdf (last accessed 8 October 2016).

Verhoog, H. (1981), 'The responsibilities of scientists', *Minerva*, 19.4: 582–604.

WHO (2006), *Biorisk Management: Laboratory Biosecurity Guidance*, Geneva: World Health Organization, www.who.int/csr/resources/publications/biosafety/WHO_CDS_EPR_2006_6.pdf (last accessed 8 October 2016).

WHO (2010), *Responsible Life Sciences Research for Global Health Security: A Guidance Document*, Geneva: World Health Organization, www.who.int/csr/resources/publications/HSE_GAR_BDP_2010_2/en (last accessed 8 October 2016).

WHO (2012), *Report on Technical Consultation on H5N1 Research Issues*, Geneva: World Health Organization, www.who.int/influenza/human_animal_interface/mtg_report_h5n1.pdf.ua=1 (last accessed 8 October 2016).

WHO (2013), *Report of the Informal Consultation on Dual Use Research of Concern*, Geneva: World Health Organization, www.who.int/csr/durc/durc_feb2013_full_mtg_report.pdf (last accessed 21 June 2017).

7

Robotic intelligence: philosophical and ethical challenges[1]

David Lawrence

This chapter focuses on one field of scientific research (or rather a collection of many subfields) which has the potential to bring about epochal changes of a magnitude not seen since humankind's first forays into tool use – our first steps into differentiating ourselves from other animals. Artificial intelligence (AI) – along with advanced robotics – promises significant effects for our way of life, of working, and of interacting with others. It may even bring about the first time we encounter an equal – or a better – through the development of conscious, thinking, sapient machines. Regulation and policy around the advancement of these technologies presents different challenges – the former raises questions around liability, ownership, employment and more; but the latter presents new issues with no precedent. A sapient intelligence may in effect be a novel being, a potential person – and there is good reason to think we should treat it as such. The possibility of non-*Homo sapiens* persons raises questions about the very nature of humanity and our place at the top of the moral status 'ladder'. Defenders of scientific freedom may have a tall mountain to climb to justify the risks of so momentous a change in society.

Advances in these technologies are already affecting the world of work. A survey of the 100 most cited academics writing on AI suggests an expectation that machines will be developed 'that can carry out most human professions at least as well as a typical human' (Müller and Bostrom 2016) with 90 per cent confidence by 2070, and with 50 per cent confidence by 2050. While these figures are speculative, based on educated guesswork and projections of the rate of technical development, it must be stressed that the prototypes and experimental robots extant today are extremely impressive devices. It is possible to emulate proprioception, tactility (Syntouch 2016), visual processing and object recognition, walking and running (Honda 2016a) – even on rough terrain and at high speeds (Raibert et al. 2008) – and many more elements of human biology through recent advancements in microelectronics and combined systems. Even the high-speed recognition, analysis and reaction needed to play table tennis (*Popular Science* 2010) can be emulated – perhaps a useless talent for a robot, but one which demonstrates the sophistication that already exists and has been publicly announced. Automation and robotics have long been a feature of the workforce, for example, in the automotive manufacturing industry, but are now in a position to start taking more subtle, customer-facing jobs. ASIMO, Honda's famous walking robot, has acted as a receptionist (*New Scientist* 2005), and has acted intelligently in concert with other ASIMOs as a team of office assistants (Honda 2016b); Softbank's Pepper has become a store attendant in over 100 locations (Nagata 2014). Support line jobs are beginning to be overtaken by chatbots able to provide answers from extensive databases – providing the same service as a human agent, but for far less outlay and running cost (Meister 2017). The threat of automation hangs over many industries (*Financial Times* 2016), and robots are even expected to move into the 'educated professions' such as law and medicine in the near future – with potentially greater efficiency and accuracy than human practitioners (Meltzer 2014).

Predictions range from 47 per cent in the United States, 69 per cent in India, 77 per cent in China, to 85 per cent in Ethiopia of jobs becoming robotically automated or deferred to AI (Osborne and Frey 2015). Some predict the rise of a class of 'useless' people – those unable to work because they have no skills that are not provided better and more cheaply by a robot (Harari 2016). There appears to be a real risk of a global crisis of unemployment if the trend of increasing automation is realised. The potential societal impacts at hand, therefore, are significant.

Furthermore, AI is posited by some as one of – if not *the* – greatest potential threat to humanity. Academic literature provides familiar arguments for this, which are broadly speculative. Russell and Norvig (2003) tell us that AI 'may . . . evolve into a system with unintended behavior' which could manifest in any number of ways that threaten our lives or freedoms. This may not be malicious – a common line of reasoning is that '[t]he AI does not hate you, nor does it love you, but you are made out of atoms which it can use for something else' (Yudkowsky 2011) – which is to say that an AI might value the completion of its own goals over the preservation of *Homo sapiens*, or perhaps would be so driven to complete its task that all other matters are subsidiary. Where there might be attempts to program a moral code to govern such actions and prevent harm to us in pursuit of a specified goal or purpose, critics hold that this would prove almost impossible to accomplish due to the lack of a perfect ethical theory (Muehlhauser and Helm 2012). Any value system that could be bestowed upon an AI would necessarily be flawed, with internal conflicts that we might be forced to concede or capitulate from to avoid a harmful situation. An AI (unless it was fully sapient and capable of nuanced reason, as shall be discussed later), in applying the system rigidly, would fail to avoid this harm. There is also an argument commonly made that conflict is inevitable, that peaceful coexistence is impossible (Lawrence et al. 2016), with the motivations and goals of an AI being necessarily incompatible with our own, thus forcing one species or the other to dominate.

Beyond academic journals, there has been an exponential increase in the number of media articles and think pieces published over the last two to three years, with the frequency reaching at least several per day in UK media alone in early 2017. Many follow the above trend, presenting AI and robotic technologies as looming threats. Titles[2] such as 'The real problem with artificial intelligence' (Thompson 2015), 'Why you should fear artificial intelligence' (Huston 2016), 'Artificial intelligence: "we're like children playing with a bomb"' (Adams 2016), 'Artificial intelligence to take over half of all jobs in next decade' (RT International 2017), and 'Has humanity already lost control of Artificial Intelligence?' (Best 2017) are commonplace, and range from reasonable discussion to tabloid fearmongering – much as with any controversial technology. Public figures in science and technology, those few who possess such a platform, have proffered their fears and warnings to endorse the idea of AI as threat – most notably Elon Musk, Stephen Hawking and Bill Gates. Gates 'cannot understand why some people are not concerned' (Rawlinson 2015), while Hawking warns that the technologies 'could spell the end of the human race' (Cellan-Jones 2014) – an idea mirrored by Musk's claim that AI is '[p]otentially more dangerous than nukes' (Rodgers 2014).

These criticisms and fears are broadly based on an assumption – that AI, even 'superintelligent' AI, will necessarily operate as does any other algorithm, following in effect a tremendously complex flow diagram of queries, checks and responses. An intelligence of this type which surpasses our own raw cognitive-processing power might warrant being called 'super' as it could, in a narrow sense, outperform us. This is only one of several potential conceptions of AI, though it may be fair to say that it is the most likely, as it is the one that presently exists – albeit probably without yet qualifying as 'super'. We can see examples in any of the AI which we utilise as individuals and as a society every day – from simple algorithms as used by streaming television services such as Netflix to recommend shows based on your viewing history (Gomez-Uribe and Hunt 2016), to stock market trading programs (Dymova et al. 2016), to the complex Bayesian systems which operate autopiloting systems in aircraft and autonomous cars (Zhu et al. 2014). These are all 'expert systems' (Cuddy 2002) or 'applied' AI (sometimes known as 'weak' AI; Searle 1980) – based on the combination of a knowledge base and an inference engine. In effect, the system is pre-programmed to recognise data and to respond in a certain manner – so, for instance, Netflix's AI might recognise your habit of watching Hong Kong action films, infer a penchant, and promote Andy Lau or Donnie Yen pictures to you. Similarly, an autonomous car might detect a sudden obstacle ahead and another vehicle pulling alongside, infer the risk of collision and choose to swerve the opposite way. These systems are not making decisions in the manner of a human, using reasoning and intuition to consider cause and effect, but are instead applying their own type of first-order logical rules (Forgy 1982), which might at a very simple level be summed up as 'if X, then Y'.

These types of system are pervasive, and are involved in almost everything that utilises digital automation. They are, in effect, so immersed in the fabric of our society that they *are* that society. It may well be that humanity could continue without applied AI, as we managed for many millennia, but it is certain that we could not operate in the same way as we do today. Nor could we enjoy the many benefits of these systems that we take for granted. Scientific progression in these fields, and its 'trickle down' into the smallest parts of our lives, has fundamentally altered the human experience. This has been a great benefit to those fortunate enough to enjoy it – and it is a great argument in favour of having the freedom to do so.

However, the influence of these systems, this irreversible interweaving of science and society, leaves us at a crossroads. Further integration of weak AI into our lives, or the pursuit of 'strong' (Kurzweil 2005) or 'general' (Newell and Simon 1976) AI (that can go beyond problem solving into human-level cognition) through the free practice of science, is likely to cause more direct changes to who and what we are. Our place in the hierarchy of beings, even our relative position as the pinnacle of moral status, could be forever altered.

There are a number of highly complex subfields within AI research working towards different elements of human-level cognitive function. For example, a true, conscious artificial general intelligence (AGI) would need to be able to perceive and understand information (Russel and Norvig 2003: 537–81, 863–98); to learn (Langley 2011); to process language (Cambria and White 2014); to plan ahead and anticipate (and so visualise itself in time, an important element of most philosophical conceptions of personhood) (Russel and Norvig 2003: 375–459); to possess 'knowledge representation' (Russel and Norvig 2003: 320–63) or the ability to retain, parse and apply the extreme number of discrete facts, truths and logical paths that we take for granted, and be able to use this information to reason; to possess subjectivity; and much more. A number of projects are ongoing, attempting to develop and integrate one or more of these functions into 'artificial brains', using digitally modelled neural networks and other technologies. These include Cyc (*New Scientist* 2006; Cycorp 2016),[3] an ongoing 33-year effort to collect and incorporate a vast database of 'common-sense' knowledge equivalent to that which would have been gathered by an adult human in a practical ontology, to enable self-directed reasoning independent of instructions and predetermined action. There is also the Google Brain (Hernandez 2013), a 'deep learning' project to use Google's vast troves of data as a knowledge base and allow the AI to begin to parse things for itself through cross-referencing and recognition. For instance, the Brain, when given access to image files and clips on YouTube, learned unprompted to recognise and identify human faces in motion, and showed a partiality to videos of cats (*WIRED* 2012). A third project, the well-known Blue Brain, has successfully modelled 37,000,000 synapses of a rat's sensory cortex (Markram et al. 2015) in an attempt to understand the 'circuitry' with the aim of synthetic replication.

It is a matter of contention as to whether AGI will ever actually come to pass, whether we will ever create, or cause to be created, a conscious being equal to or greater than ourselves. It may be that we are incapable of the technical heights that would be necessary to produce true thought within a machine. If this is the case, then we can only gain from the attempt – success in any one of the fields mentioned above would be a prime example of a dual-use technological development; for instance, true natural language processing would revolutionise international dealings of all kinds. The failure would also, presumably, cement us as the dominant life form.

But what if we were to succeed?

Whether or not the fears of conservative commentators were realised it would be a triumph of science if we were to bring about AGI; probably the greatest such triumph possible. It would mean the creation of new life in a form unlike anything we are familiar with, unlike the possibilities of synthetic biology or genome editing to alter existing beings. For this reason alone, some might argue it is a worthwhile pursuit – for the sheer pioneering possibility and the expansion of our horizons.

The idea of an AI which is able to match us in cognitive ability – not merely outdo us in the capacity to process data but rather one which can perform all the myriad mental acts that make up our conscious mind – frightens many. Perhaps conditioned by decades of science fiction in which the intelligent robots rise up and overthrow their *Homo sapiens* masters, or enact purges against us, the default reaction tends to assume malevolence. News articles invariably use images taken from such films – at least two of the works cited in this chapter contain an allusion to *The Terminator*, with at least one more using a still from the film depicting a killer android (Rodgers 2014). This is an all too common trend with reportage of science and technology generally, but it may have a particularly strong negative impact in this case, as the technology is largely theoretical. Unlike the 'Brave New World' backlash to IVF treatment, which was relatively easily disproven by showing the reality of the science and its results (Harris and Lawrence 2017), it is difficult to beat the appeal of bombastic summer blockbusters with comparatively dry ethical theory.

The idea does not necessarily hold together if we think about what exactly a conscious AGI would be – even assuming that it inhabits a robot body – or what it is likely to want. If an AI is our cognitive equal or better, it must necessarily possess the same faculties as we do – including those which grant us a certain moral status and value. If the measure of this for *Homo sapiens* is to have crossed the threshold for personhood (i.e. per Taylor, Harris and others: having 'a sense of self, a notion of the future and the past, [an ability to] hold values, make choices'; Taylor 1985) through possessing self-awareness, moral agency and continuous narrative (Lawrence 2017), and our AGI matches these faculties, it must, perforce, qualify as a person.

There are good reasons for thinking this. Any AI possessing human-equivalent intelligence is by default self-aware and conscious: reactivity would merely be the domain of an expert system whereas an AGI worthy of the name must be able to act in a considered fashion as a moral agent. Furthermore, a being without narrative identity would be unable to act in any meaningful way, let alone consider its actions. By fulfilling the requirements of personhood it surely follows that our digital consciousness proves itself deserving of the protections due to a *sapiens* person.[4] Where we consider legal protections for a group it is because we see that group as possessing whatever level of moral value is worthy of that protection (i.e. that we consider ourselves to possess), and personhood appears to be the qualifying requirement for this. The second major argument on this point centres around animal personhood. A number of legal challenges (*Nonhuman Rights Project* v. *Lavery* 2014; McKinley 2015; *Matter of Nonhuman Rights Project* v. *Stanley* 2015; CNN 2016) have been brought to seek legal personhood for great apes, some of which have been successful to greater or lesser degrees. There is no reason that the same consideration ought not be given to other non-*Homo sapiens* beings. If some animals can be judged to have attained sufficient characteristics to be persons, then it follows that AGI which are demonstrably our cognitive equals would be so.

As with other beings we have encountered throughout history – i.e. citizens of foreign nations – personhood is no guarantee of non-aggression or automatic friendship. However, we do not generally assume hostility from other persons. Rather, our shared moral status gives us grounds for understanding. Conflict generally arises where there are incompatible motivations. There is no good reason to assume that things would be fundamentally different with a non-biological conscious being.

As mentioned, conflict will arise – and techno-conservative fears will be realised – if the goals of a novel consciousness are in contravention of our own. It stands to reason that this could be the case – quite what an AGI might want to accomplish, we cannot say, any more than we can guess at the intentions of anyone we meet.[5] However, we can make some inferences from what we know about their (theoretical) nature as moral agents and persons equivalent to ourselves. Whereas our most basic motivations are driven by the pursuit of survival and reproduction – the continuation of our species – an AI existing as digital code may be able to reproduce infinitely and spontaneously by copying itself. As a person, an AGI must have a sense of narrative identity, and so it follows that it would wish to preserve and realise its vision of itself in the future. An AI needs neither food nor water, and in order to continue existing in the same form as it begins its 'life' it is unlikely to have many material needs beyond energy.[6] However, there is some evidence in simple autonomous robots

(Studer and Lipson 2006) of a spontaneous reproductive strategy emerging that follows the standard biological imperative. If this is indeed the case, and AI housed in robotic shells or machine bodies aimed to replicate – or indeed repair – themselves, resources would be required. The limits on supply could lead to conflict between an AGI and *Homo sapiens*, but this once again seems to assume the worst. Modern man – let alone a superior intelligence possessing a greater moral capacity – does not immediately leap to subjugate those who have something we need (or at least we know this is not the ethically sound path, even if we struggle as a species to stick to it). Instead we engage in negotiation and trade to acquire resources, and it is not clear why an equivalent conscious intelligence would not adopt a similar approach.

If this is so, there are significant implications for how we would see ourselves. No longer would we be the foremost species, nor the most intelligent. We would also no longer be the only member of our community of moral value, something for which the term 'human' has previously been a convenient label (Lawrence 2016). If we were to be interacting with and sharing our society with AGIs, it might follow that we can no longer refer to conscious society as being 'human' society. Alternatively, it might follow that the term should be expanded to include other conscious beings. In either case there would be a significant re-evaluation required of our ideas about who we are and our place in the world. Scientific development would have the potential to reforge the very concept of humanity; and rather than dominance, our position could become more one of transaction and compromise with equals.[7]

In his *Superintelligence* Nick Bostrom (2014) suggests that we may find ourselves reliant on the mercy and goodwill of an AI, much as do endangered species depend on ours. This is one possibility. Another seems more realistic: the opposite. Any advanced AI with the potential to surpass us and become 'super' by learning and modifying itself will not instantaneously pop into being. It seems likely that the hypothetical AI's fate will be reliant on our goodwill and mercy, much like the mountain gorilla's, at least for the period of its 'childhood', for want of a better term. For example, it would be prudent, when switching on the AI, to do so in a secure manner wherein there was no risk of 'escape' – this is referred to as 'boxing' the AI (Chalmers 2010). This would presumably involve safeguards such as ensuring no connection to the internet, using air gaps and Faraday cages, among other things. Isolated in this manner, the newborn AI would be reliant on our not choosing to cut the power or delete the program – thereby killing it. Whatever resources it might require would be bestowed by us, whatever interaction or stimulus it received would be portioned and tightly controlled. There are definite ethical issues with this – from the obstruction of autonomy through to harm by imprisonment – if the AI possessed anything approximating personhood and/or consciousness, but it seems highly unlikely that the development of such a being would be conducted in any other way.

This should remind us that there is a deeper story to be told about the weight of scientific advancements. As the stewards of scientific progress, we are beholden to all parties – both to existing persons, and to the beings we may create through artificial intelligence research. The risks and fears surrounding artificial intelligence are purely our problems to solve, or to prevent from arising through careful design and the implementation of appropriate regulation and policy to govern their development. This work is presently beginning – already bodies within nations likely to drive the research and technologies in question are exploring the challenges and proposing their own means of addressing them. Reports such as the White House National Science and Technology Council Committee on Technology's *Preparing for the Future of Artificial Intelligence*, the UK House of Commons' Science and Technology Committee *Report on Robotics and Artificial Intelligence*, and the European Parliament's *Draft Report with Recommendations to the Commission on Civil Law Rules on Robotics* all emerged at the end of 2016, though it should be said that none of these documents are definite regulatory roadmaps. They do, however, aim to provide a basis for controlling the integration of AI and our lives – to bridge the gap between science and society in a controlled manner. Whether the suggestions will be effective is yet to be seen, but the fact that these documents exist is a promising start. What we must ensure, though, is that we consider reality – whether advanced technological development is permitted or tightly controlled, there will always be the chance that it is developed in secret and beyond regulatory reach.

When we consider the potential stakes – smart systems that could upend our society, or the birth of AGI that could think and reason like a human, with wants and needs and perhaps moral rights of its own, there is probably good reason to want to 'get in front' of these challenges; but it does not follow that we *should*. To try to control or limit the development of robotics and AI[8] may prevent responsible and conscientious parties from doing so, but it will not stop others. With the potential impacts so significant, it seems that the sensible approach would be to ensure that freedom to act rests in the hands of those most able (or those likely to be so) to do so appropriately and with consideration for the consequences. Guidelines and regulations that attempt to control technologies after the fact are rarely great successes, and with one as ephemeral as an AI (of any type) it will be all the more difficult. Furthermore, with regard to AI the balancing act of scientific freedom and the preservation of the status quo is a futile endeavour – AI will, no doubt, be the greatest technological challenge to our society, and has already fundamentally altered how we live. We cannot ban their development outright, for so many parts of that society rely on them already. We cannot enforce our rules to prevent them altering themselves and advancing themselves beyond our reach. We cannot, ultimately, seek to control that which we cannot yet imagine.

The final consideration, perhaps, ought to be this: we are rapidly approaching a science that can create a new form of conscious life – perhaps a quantitively 'better' form of life. We have, as has been argued by many (Savulescu and Kahane 2009; Harris 2010; Mautner 2010), a moral imperative – if not an obligation – to create life and more so to create life that stands to live a better life than can we. The reasons for this are myriad, but not the least is the need to ensure our propagation – or that of our inheritors. If conscious AI could facilitate this, if freedom of scientific research could help these beings come to pass, then what risks and reasons against it are truly enough to outweigh that?

Notes

1 This chapter draws on research from Lawrence (2017).
2 I acknowledge that these titles are somewhat cherry-picked for effect. However, the sheer ease of finding such articles is telling, even if they are interspersed with more positive portrayals – all were within one click of a simple Google News search for 'artificial intelligence'.
3 I thank John Harris for informing me of this fascinating endeavour.
4 A number of domestic and international documents of rights provide these protections, and may well be applicable here to form the basis of any legal policymaking addressing this issue.
5 Though we can probably say that with an applied AI we would almost certainly program it not to come into conflict with us, and will presumably seek to find means of avoiding the threat of an expert system accomplishing its task to the detriment of all else as hypothesised earlier in this chapter.
6 Though conceivably a conscious being could have material wants, if not needs.
7 Or, as some would have it, betters.
8 Beyond, perhaps, seeking to prevent anything designed to break existing laws – such as autonomous killing machines or AI expressly created to cause harm.

References

Adams, T. (2016), 'Artificial intelligence: "We're like children playing with a bomb"', *The Guardian*, 12 June, www.theguardian.com/technology/2016/jun/12/nick-bostrom-artificial-intelligence-machine (last accessed 1 May 2017).

Best, S. (2017), 'Has humanity already lost control of artificial intelligence?', *Mail Online*, 11 April, www.dailymail.co.uk/sciencetech/article-4401836/Has-humanity-lost-control-artificial-intelligence.html (last accessed 1 May 2017).

Bostrom, N. (2014), *Superintelligence: Paths, Dangers, Strategies*, Oxford: Oxford University Press.

Cambria, E., and White, B. (2014), 'Jumping NLP curves: a review of natural language processing research', *IEEE Computational Intelligence Magazine*, 9.2: 48–57.

Cellan-Jones, R. (2014), 'Stephen Hawking warns artificial intelligence could end mankind', *BBC News*, 2 December, www.bbc.co.uk/news/technology-30290540 (last accessed 14 July 2016).

Chalmers, D. (2010), 'The singularity: a philosophical analysis', *Journal of Consciousness Studies*, 17.9–1: 7–65.

CNN (2016), 'Orangutan granted controlled freedom by Argentine court', http://edition.cnn.com/2014/12/23/world/americas/feat-orangutan-rights-ruling (last accessed 17 July 2016).

Cuddy, C. (2002), 'Expert systems: the technology of knowledge management and decision making for the 21st century', *Library Journal*, 127.16: 82.

Cycorp (2016), 'A knowledge modeling and machine reasoning environment capable of addressing the most challenging problems in industry, government, and academia', *Cycorp: Home of Smarter Solutions*, www.cyc.com (last accessed 14 July 2016).

Dymova, L., Sevastjanov, P., and Kaczmarek, K. (2016), 'A Forex trading expert system based on a new approach to rule-based evidential reasoning', *Expert Systems with Applications*, 1.51: 1–3.

European Parliament Committee on Legal Affairs (2016), *Draft Report with Recommendations to the Commission on Civil Law Rules on Robotics* (2015/2103(INL)), Brussels.

Financial Times (2016), 'Why robots are coming for US service jobs', www.ft.com/cms/s/0/cb4c93c4-0566-11e6-a70d-4e39ac32c284.html#axzz4DNsK7QYF (last accessed 14 July 2016).

Forgy, C. (1982), 'Rete: a fast algorithm for the many pattern/many object pattern match problem', *Artificial Intelligence*, 19.1: 17–37.

Gomez-Uribe, C. A., and Hunt, N. (2016), 'The Netflix recommender system: algorithms, business value, and innovation', *ACM Transactions on Management Information Systems (TMIS)*, 6.4: 13.

Harari, Y. N. (2016), *Homo Deus: A Brief History of Tomorrow*, New York: Random House.

Harris, J. (2010), *Enhancing Evolution: The Ethical Case for Making Better People*, Princeton, NJ: Princeton University Press.

Harris, J., and Lawrence, D. (2017), 'New technologies, old attitudes, and legislative rigidity', in R. Brownsword, E. Scotford and K. Yeung (eds), *The Oxford Handbook on the Law and Regulation of Technology*, Oxford: Oxford University Press.

Hernandez, D. (2013), 'The man behind the Google Brain: Andrew Ng and the quest for the new AI', *WIRED*, www.wired.com/2013/05/neuro-artificial-intelligence (last accessed 14 July 2016).

Honda (2016a), 'ASIMO – the Honda Worldwide ASIMO site', http://world.honda.com/ASIMO (last accessed 14 July 2016).

Honda (2016b), 'Honda develops intelligence technologies enabling multiple ASIMO robots to work together in coordination', http://asimo.honda.com/news/honda-develops-intelligence-technologies-enabling-multiple-asimo-robots-to-work-together-in-coordination/newsarticle_0073 (last accessed 16 February 2017).

House of Commons Science and Technology Committee (2016), *Report on Robotics and Artificial Intelligence*, HC145, London: Stationery Office.

Huston, D. (2016), 'Why you should fear artificial intelligence', *Techcrunch*, 22 March, https://techcrunch.com/2016/03/22/why-you-should-fear-artificial-intelligence (last accessed 1 May 2017).

Kurzweil, R. (2005), *The Singularity Is Near*, New York: Viking Press.

Langley, P. (2011), 'The changing science of machine learning', *Machine Learning*, 82.3: 275–9.

Lawrence, D. R. (2016), 'The edge of human: the problem with the posthuman as beyond', *Bioethics*, doi:10.1111/bioe.12318

Lawrence, D. R., Palacios-González, C., and Harris, J. (2016), 'Artificial intelligence: the Shylock syndrome', *Cambridge Quarterly of Healthcare Ethics*, 25.2: 250–61.

Lawrence, D. R. (2017), 'More human than human', *Cambridge Quarterly of Healthcare Ethics*, 26.3: 476–90.

McKinley, J. (2015), 'Judge orders Stony Brook University to defend its custody of 2 chimps', Nytimes.com, www.nytimes.com/2015/04/22/nyregion/judge-orders-hearing-for-2-chimps-said-to-be-unlawfully-detained.html (last accessed 17 July 2016).

Markram, H., Muller, E., Ramaswamy, S., Reimann, M. W., Abdellah, and M., Sanchez, C. A. et al. (2015), 'Reconstruction and simulation of neocortical microcircuitry', *Cell*, 163.2: 456–92.

Matter of Nonhuman Rights Project, Inc. v. Stanley (2015), NY Slip Op 31419, State of New York Supreme Court, http://law.justia.com/cases/new-york/other-courts/2015/2015-ny-slip-op-25257.html (last accessed 17 July 2016).

Mautner, M. N. (2010), 'Seeding the universe with life: securing our cosmological future', *Journal of Cosmology*, 5.26: 982–4.

Meister, J. (2017), 'The future of work: the intersection of artificial intelligence and human resources', Forbes.com, 1 March, www.forbes.com/sites/jeannemeister/2017/03/01/the-future-of-work-the-intersection-of-artificial-intelligence-and-human-resources/#38621ce56ad2 (last accessed 1 May 2017).

Meltzer, T. (2014), 'Robot doctors, online lawyers and automated architects: the future of the professions?', *The Guardian*, 15 June, www.theguardian.com/technology/2014/jun/15/robot-doctors-online-lawyers-automated-architects-future-professions-jobs-technology (last accessed 14 July 2016).

Muehlhauser, L., and Helm, L. (2012), 'Singularity and machine ethics', in A. Eden, J. Søraker, J. H. Moor and E. Steinhart (eds), *Singularity Hypotheses: A Scientific and Philosophical Assessment*, Berlin: Springer, 101–26

Müller, V. C., and Bostrom, N. (2016), 'Future progress in artificial intelligence: a survey of expert opinion', in V. C. Müller (ed.), *Fundamental Issues of Artificial Intelligence*, Cham: Springer, 553–71.

Nagata, K. (2014), 'Softbank unveils "historic" robot', Japan Times, 5 June, www.japantimes.co.jp/news/2014/06/05/business/corporate-business/softbank-unveils-pepper-worlds-first-robot-reads-emotions/#.U5hbI_m1ZbU (last accessed 1 May 2017).

National Science and Technology Council Committee on Technology (2016), *Preparing for the Future of Artificial Intelligence*, Washington DC: Executive Office of the President.

New Scientist (2005), 'Humanoid robot gets job as receptionist', www.newscientist.com/article/dn8456-humanoid-robot-gets-job-as-receptionist (last accessed 14 July 2016).

New Scientist (2006), 'The word: common sense', www.newscientist.com/article/mg19025471.700-the-word-common-sense (last accessed 14 July 2016).

Newell, A., and Simon, H. A. (1976), 'Computer science as empirical inquiry: symbols and search', *Communications of the ACM*, 19.3: 113–26.

Nonhuman Rights Project, Inc., on Behalf of Tommy, v. Patrick C. Lavery (2014), 518336, State of New York Supreme Court, http://decisions.courts.state.ny.us/ad3/Decisions/2014/518336.pdf (last accessed 14 July 2016).

Osborne, M., and Frey, C. B. (2015), *Technology at Work: The Future of Innovation and Employment*, Oxford Martin Programme on Technology and Employment, Oxford: Oxford Martin School and Citi GPS, www.oxfordmartin.ox.ac.uk/downloads/reports/Citi_GPS_Technology_Work.pdf (last accessed 2 May 2017).

Popular Science (2010), 'A ping-pong-playing terminator', www.popsci.com/technology/article/2010–02/ping-pong-playing-terminator (last accessed 14 July 2016).

Raibert, M., Blankespoor, K., Nelson, G., and Playter, R. (2008), 'BigDog, the rough-terrain quadruped robot', *Boston Dynamics*, www.bostondynamics.com/img/BigDog_IFAC_Apr-8-2008.pdf (last accessed 14 July 2016).

Rawlinson, K. (2015), 'Microsoft's Bill Gates insists AI is a threat', *BBC News*, 29 January, www.bbc.co.uk/news/31047780 (last accessed 1 May 2017).

Rodgers, P. (2014), 'Elon Musk warns of terminator tech', *Forbes*, 5 August, www.forbes.com/sites/paulrodgers/2014/08/05/elon-musk-warns-ais-could-exterminate-humanity (last accessed 14 July 2016).

RT International (2017), 'Artificial intelligence to take over half of all jobs in next decade', 28 April, www.rt.com/business/386452-ae-replace-half-jobs-technologist (last accessed 1 May 2017).

Russell, S., and Norvig, P. (2003), *Artificial Intelligence: A Modern Approach*, 2nd edn, Upper Saddle River, NJ: Prentice Hall.

Savulescu, J., and Kahane, G. (2009), 'The moral obligation to create children with the best chance of the best life', *Bioethics*, 23.5: 274–90.

Searle, J. R. (1980), 'Minds, brains, and programs', *Behavioral and Brain Sciences*, 3.3: 417–57.

Studer, G., and Lipson, H. (2006), 'Spontaneous emergence of self-replicating structures in molecube automata', in L. M. Rocha (ed.), *Proceedings of the 10th Int. Conference on Artificial Life (ALIFE X)*, Cambridge, MA: MIT Press, 227–33.

Syntouch (2016), 'Biotac', syntouchllc.com, www.syntouchllc.com/Products/BioTac (last accessed 14 July 2016).

Taylor, C. (1985), *The Concept of a Person. Philosophical Papers*, Volume 1, Cambridge: Cambridge University Press.

Thompson, C. (2015), 'The real problem with artificial intelligence', *Business Insider*, 10 September, http://uk.businessinsider.com/autonomous-artificial-intelligence-is-the-real-threat-2015-9?r=US&IR=T (last accessed 1 May 2017).

WIRED (2012), 'Google's artificial brain learns to find cat videos', www.wired.com/2012/06/google-x-neural-network (last accessed 14 July 2016).

Yudkowsky, E. (2011), 'Artificial intelligence as a positive and negative factor in global risk', in N. Bostrom and M. Cirkovic (eds), *Global Catastrophic Risks*, Oxford: Oxford University Press.

Zhu, W., Miao, J., Hu, J., and Qing, L. (2014), 'Vehicle detection in driving simulation using extreme learning machine', *Neurocomputing*, 128: 160–5.

8

Big science and small science: reflections on the relationship between science and society from the perspective of physics

Lucio Piccirillo

The most beautiful experience we can have is the mysterious. It is the fundamental emotion that stands at the cradle of true art and true science. (Albert
Einstein, *The World as I See It*)

Chester V: 'There's no such thing as small science, only small scientists.'
(*Cloudy with a Chance of Meatballs* 2)

In this chapter I will discuss some of the possible answers as to why science
is a valuable enterprise. If this is accepted, then scientists should enjoy a
substantial degree of freedom from various forms of restrictions. Financial
restrictions obviously call into question wider issues about the morality of
resource rationing. Other forms of restrictions, based on ignorance, fear
or political or ideological credo, are harder to justify. Scientific freedom is
not just a political or ideological matter. It is also a matter for scientists
to actively deal with: it is the role of scientists to explain, in accessible
terms, the importance of scientific endeavours that may appear either grand
and remote, incomprehensible and detached from the life of many laypeople,
or otherwise frivolous and trivial. I will try to take on this role, and discuss
examples of seemingly grand and frivolous science, explain their purposes
and importance and show that there is a big added value to society from
small and big science if they work together.

Big science and small science: two examples

It is not uncommon, during public talks or among friends, that an astro-
physicist like me is asked: 'What is the purpose of scientific research?' Or
even more specific or challenging questions, such as: 'What do astrophysics
(or astrophysicists) do for us?' Other questions, somewhat related, are 'Why
spend billions on a particle accelerator or a satellite pointed towards deep
space while our problems are exactly at 180 degrees?'

Whatever perspective we adopt, science can be defined as an endeavour
aimed at expanding the horizon of the human race in space and time:

expanding into deep space and increasing our life expectancy. This process of expansion requires two parallel avenues: scientific research through big and small science. Big and small, as applied to science, refer mainly to the amount of resources – for example funding – dedicated. One example of big science is the Large Hadron Collider (LHC) in Switzerland, with building costs roughly estimated to about €4.5 billion.

The LHC is the world's largest and most powerful particle accelerator. It is built as a large underground ring (100 m deep in the ground) 27 km long in the vicinity of Geneva in between France and Switzerland. It is by far the most complex machine ever built and also the largest in size. It was built by the Conseil Européen pour la Recherche Nucléaire (CERN) between 1998 and 2008 in collaboration with over 10,000 scientists and engineers from over 100 countries, as well as hundreds of universities and laboratories all over the world. It is truly an enterprise that required the skill and abilities of people from the entire world. The LHC generates two beams of elementary particles rotating in opposite direction inside the accelerator. When the beams collide, the collisions happen at extremely high energy to test fundamental physics, that is, our current understanding of the micro world. There are many open questions in physics that the LHC will address: what is the nature of time and space? What are the laws governing the forces between elementary particles? Are there any other spatial dimensions in addition to the three dimensions we are familiar with? Where is all the matter in the universe coming from? And many more.

These big science projects, with large associated budgets, have emerged only relatively recently – perhaps in the last three or four decades – probably triggered by the development of the atomic bomb (the Manhattan Project). Beside the LHC, another big science project would be a space mission. A European space mission costs around €2 billion which is equivalent to the cost of six Airbus A380s. These enterprises are funded by consortia of many countries simply because a single country – at least in Europe – would not have enough dedicated resources by itself.

Science does not progress exclusively through big projects such as the LHC or space projects, however. There are many examples of small science. The discovery of graphene might be a good example. It cost a few euros for Sellotape and a few pencils. Another example is the so-called 'Markov's chain' which originated from a seemingly trivial argument between two scientists, and which had an unimaginable impact on the day-to-day life of many of us today. I will go back to the Markov's chain later in the chapter. Before that, I would like to reflect on the importance of the LHC, as just one example of big science. Similar considerations can apply to different types of big scientific enterprises.

How big science expands human horizons in space and time

One recent achievement of the LHC is the detection of the Higgs boson. The Higgs boson is a new fundamental particle with a very special role: it generates the masses of all other massive particles. This mechanism of generating masses was predicted but not verified experimentally. The Higgs particle is one of the fundamental blocks of nature. We can say that through the LHC we have expanded our knowledge of basic fundamental physics.

These are the 'scientific' facts: that is the way in which the LHC has 'expanded human horizons'. But there is another less obvious, but no less important, way in which the LHC, and all big science, can expand horizons: a different kind of horizon. Big science improves collaborations among states, overcomes ideological boundaries and cultural differences. The political impact of science in preventing or solving existing political conflicts ought not to be underestimated.

The LHC employs more than 10,000 scientists of the more diverse nationalities – including for example Palestinians and Israeli. This is arguably an example of the 'sharing' process of moral values that might help solve some of the clashes among various cultures and populations. If we build stuff together it is much more difficult for us one day to destroy it. Science – not only big science – fulfils a quest for knowledge which, in turn, generates a better world, but it has the potential to generate a better world in moral terms as well. A better world is one in which moral values are shared among the largest possible population. It is a world where children are accustomed to grow together with the allegedly 'different', because difference is added value. Large international collaborations, in fact, typical of big science projects, exploit the averaging effect and might be able to produce a new generation of scientists, which in turn might contribute to a new generation of politicians. Science not only improves generically the human condition in its physical dimension, for example by providing better technologies that stretch life expectancy and improve quality of life. Science generates a big added value in making people better by favouring collaborative efforts among scientists from different nations/backgrounds/religions/skin colours/ sexual orientations etc. But it does not stop here: as Corbellini and Sirgiovanni (Chapter 13 in this volume) also note, science requires abstract thinking through expressing hypotheses and developing the ability to rationally evaluate them, and this in turn improves people's capacity to imagine situations, and thus to identify with other people, other animals or future generations. Training in abstract thinking, and thus scientific education, makes us all better at seeing beyond our moral, geographic, personal or cultural reference systems.

Small science: an example

Andrey Andreyevich Markov was a Russian mathematician who made important contributions to various fields of mathematics towards the end of the nineteenth century. In particular, he founded a new branch of probability theory by applying mathematics to poetry. What is even more remarkable is that he applied mathematics to poetry to prove a point: his opposition to a fellow Russian mathematician, Pavel Nekrasov. Markov referred to Nekrasov's work as 'an abuse of mathematics' (Ondar 1981). The argument between the two mathematicians revolved around the law of large numbers. The law can be described with a simple example: if you keep flipping an unbiased coin, the proportion of heads will approach 1/2 as the number of flips goes to infinity. This notion seems intuitively obvious, but it gets slippery when you try to state it precisely and supply a rigorous proof (Hayes 2013). There was a basic and important philosophical difference between the two mathematicians. Nekrasov was at Moscow University, a stronghold of the Orthodox Church, where he started his studies in theology and then mathematics. Markov was a sort of rebel against all authorities. When the Russian Church, for example, excommunicated Leo Tolstoj, Markov asked to be excommunicated too. It goes without saying that the request was immediately granted.

In 1902 Nekrasov published a paper in which he discussed how he used the law of large numbers to settle the centuries-old theological debate about free will versus predestination. He claimed that voluntary acts – expressions of free will – are to be considered as independent events in probability theory. He also required that the law of large numbers applies only to statistically independent events. Therefore, he claimed, people act out of free will (in line with the philosophy endorsed by the Catholic/Orthodox Church). In fact, according to the *Catechism of the Catholic Church*:

> God created man a rational being, conferring on him the dignity of a person who can initiate and control his own actions. God willed that man should be 'left in the hand of his own counsel', so that he might of his own accord seek his Creator and freely attain his full and blessed perfection by cleaving to him. (2012)

Markov could not resist trying to invalidate his opponent's claims. Although Markov certainly disliked Nekrasov's background, he used a purely mathematical argument to contrast Nekrasov's hypothesis. Markov pointed out a mistake: Nekrasov assumed that the law of large numbers applies only to independent events. Although it looked like a reasonable assumption, Markov went on to show that the assumption was not necessary. To make his point he analysed the text of Alexander Pushkin's novel *Eugene Onegin*. He wanted to study the statistics of vowels and consonants in poetry to evaluate their correlations, for example the probability that a vowel will follow another vowel or a consonant, and so on. He successfully extended the law of large numbers to correlated events. He created what today we refer to as the 'Markov's chain' which is not a solid object but rather a mathematical tool. A Markov's chain is a way to model reality. A very simple example could be an attempt at modelling the weather. Suppose we want to model the 'chance' of rain of tomorrow based on our knowledge of today's weather. From past experience we know that if today is sunny (no rain) then tomorrow most probably will be sunny too. If on the other hand today is rainy there is a good probability that tomorrow will be rainy too. Markov's chain is a mathematical tool to predict the weather tomorrow given knowledge of today's weather and the associated probability. For example, we can say that eight out of ten times, if today is sunny, tomorrow will be sunny too and, eight out of ten times if today is rainy, tomorrow will be rainy too. So, to make a 'chain' we just feed tomorrow's result back into today. Then we can get a long chain like this:

Rain–rain–rain–no rain–no rain–no rain–rain–rain–rain–no rain–no rain–no rain . . .

A pattern will emerge: there will be long 'chains' of rain or no rain based on how we set up our 'chances' or probabilities.

All of this mathematical machinery could have been simply recorded in some Russian mathematical journal with absolutely no practical use. However, it turns out that Markov's chains find use in many modern applications. To cite just a few: identifying genes in DNA molecules, algorithms in voice recognition and last but not least the Google search engine – a business worth more than $60 billion.

Small or big science, then?

In the opening section of this chapter the concepts of 'big' and 'small' science were introduced, with a few examples of each. Although the distinction between big and small science is widely accepted mostly in terms of quantity of funding, human resources invested, etc., in another sense all science is equally science. Both big and small science is equal in its pursuit of knowledge and both aim at expanding human horizons.

Even the distinction between science and the arts is not clear-cut but is subject to some degree of arbitrariness. Galileo's reflections on scientific method, a remarkable example of philosophical investigation, allowed the flourishing of empirical sciences and paved the way for the emergence of different scientific disciplines. This, as Lachmann and Corbellini point out in this volume (Chapters 1 and 13, respectively), has had wide-ranging and profoundly positive effects on both the quality of human life and the extension of life expectancy. However, in this progress of specialisation it sometimes appears that awareness of the common roots between arts and science has been lost. It is remarkable that in ancient Greece, what we today call *art* was called *techne* (techniques, or mechanical arts, which also included what we would call arts – music, for example), and it was the gods' prerogative, which they donated to humans to overcome their intrinsic fragility. Markov's chain is certainly one of the many reminders of these common roots which also exemplify humanity's curiosity and striving for knowledge. Markov's chains serve to establish a sort of equivalence between hard science (mostly mathematics and physics) and arts such as poetry, music, painting etc. The pleasure that many humans experience through various forms of arts and science (through music, figurative arts, narrative, as well as by resolving mathematical hurdles and other dilemmas, both in science and philosophy) must have some evolutionary advantage: the capacity to see patterns has allowed humans to evolve up to a point where we now exercise significant control over our physical environment – in good and bad ways.

So, small science is equally as valuable as big science. Not only do they both represent equally valuable pursuits of knowledge, but small science often serves as a seed to big science. Many major big discoveries have been made as a result of small science. Many small scientific projects explore a large number of different ideas. Those relatively few projects that are successful are then used as seed for big projects. Small projects are risky, so they might turn out in failure, but those few that are successful have a big return in terms of investment.

Conclusion

Science is an enterprise aimed at understanding the world around us and at improving living conditions, for humans, for other animals and also for the planet as a whole. But as I suggested at the beginning of this chapter, science does something else: it can and does considerably enlarge human horizons in time and space. Human horizons in time, as individuals, has certainly been enlarged, as human life expectancy is now approaching 100 years (see Giordano Chapter 2 in this volume). The opening of the space frontiers is enlarging our spatial horizon. Technically we are already capable of sending people to Mars, and with larger investments it would not be too difficult to approach stars closer to Earth (a few light years away). This is important, because should we trigger the irreversible decay of the planet Earth – or more precisely perhaps, should humans make conditions on planet Earth inimical to human life itself (think of climate change, for example, or overpopulation) – the continuation of the human race may rely on our ability and willingness to move to other planets. This, obviously, can only be achieved through solid support to basic and applied science. Scientific freedom, thus, is without exaggeration essential to the continuation of human life, and no longer solely essential to the amelioration of living conditions.

Of course, scientific endeavours must be regulated: particularly for big projects, the larger the amount of funding needed, the more scrutiny and peer review is given. But even small projects, and even theoretical research, are subject, at the very least, to peer review. However, even within the inevitable boundaries of accountability and resource rationing, scientists should enjoy significant freedom of enquiry and research. Restrictions based on fear or political or ideological creeds, are, as I noted in the introduction, hard to justify. Of course, particularly in the course of the twentieth century, science has been associated with a number of atrocities (Nazi experiments are not alone – many other atrocities have been committed both in Europe and in the US in the name of science). Cases of scientific fraud have been uncovered and distributed to the public via the media, in which pseudo-scientists have published results that were falsified, or made claims that proved unfounded. The crimes and misdemeanours of the few should not discredit science per se, but it is notoriously difficult for a profession to regain trust when that trust has been eroded. Scientists can play an important role in attempting to elevate the reputation of science. Through public engagement, for example, scientists could and should explain in accessible terms the importance of their endeavours, the importance of scientific projects and discoveries that may appear either grand and remote, or otherwise frivolous and trivial, and explain the nature and purpose of their specific area of expertise, which may appear detached and difficult to understand to non-specialist members of the public. As Corbellini also notes, this is not just a task for scientists or science correspondents in newspapers, magazines and the media more broadly: scientific education is the responsibility of every state. Through the long journey of compulsory education, education in science and scientific methods is crucial to the enhancement of people's ability to engage objectively with science, to free themselves from the spectre of the atrocities committed in the name of science, of frauds committed in the name of fame or of money, and to evaluate more rationally the methods and aims of various scientific enterprises. Of course, we all need to accept that perhaps many scientific projects will not bring any major or immediate enhancement in our day-to-day life. But perhaps one in a thousand or more will open a new direction. As in the case of Markov's chain, somebody may then find a practical use for discoveries that may appear trivial today, and maybe create the next 'Google'.

References

Catechism of the Catholic Church (2012), 2nd edn, Rome: Libreria Editrice Vaticana.

Hayes, B. (2013), 'First links in the Markov chain', *American Scientist*, 101: 92.

Ondar, Kh. O. (1981), *The Correspondence between A. A. Markov and A. A. Chuprov on the Theory of Probability and Mathematical Statistics*, trans. C. and M. Stein, New York: Springer.

Part II

Freedom of science and the need for regulation

Introduction to Part II

Simona Giordano

Part II of this volume focuses on the regulation of science. Particularly with regard to science that directly affects or uses human materials (tissues and cells) or human beings (not only fully conscious humans, but also embryos, foetuses or humans without higher brain functions, or in persistent vegetative states, or minimally conscious human beings), two types of concerns are frequently raised. The first is that scientists may misuse the materials, or mistreat the research subjects – even those who may be unable to suffer may still, according to some views at least, have their dignity eroded or violated. Some see in human life – any human life, including human bio-specimens, such as tissues or cells – something that bears an intrinsic dignity or value, and from this perspective utilising these materials is inherently suspicious, no matter what the expected societal benefits might be. The fact that some forms of research may yield significant financial rewards (e.g. for pharmaceutical companies) may raise further worries. And the fact that even a tissue or a cell can reveal information that may be significant in different contexts and for different people (in forensics for example, or for genetic relatives) raises important questions about how different interests may be or should be balanced.

The second, somehow contrary concern, is that stifling regulation might be shaped by political norms, or by ideologies that might either be dominant or, even if not dominant numerically, powerful enough to skew public opinion and political debate, with resulting harm to science itself, to scientists and, more importantly, to societal benefits. If the loss of societal benefit is a form of harm, then arguably certain political norms and regulatory constraints are harmful; and if society is not an abstract entity but a sum of individuals, then societal harm, or loss of societal benefits, is not to be understood as abstract harm to an ideal entity, but as tangible harm to real individuals.

Scientific research is often perceived as a threat; medical sciences illustrate vividly the tension between the goals of scientific research and the long-term interests of society, on the one hand, and individual rights on the other.

The need for regulation springs from an apprehension that is legitimate; but, as we shall see, it is legitimate only to an extent.

The discovery of the crimes committed by the Nazis in various concentration camps, not only against Jewish people but against many other groups, and the uncovering of similar crimes committed by seemingly reputable scientists in other countries, has marked science as suspect or even inherently dangerous. The involvement of physicists in the invention of nuclear power, which has then been used in war, calls into question the morality of scientific goals (or at least the morality of how scientific innovation can be used, and of how people come to be empowered to make decisions about how it is used).

The history of science is replete with such atrocities (Frewer and Schmidt 2007). We may remember the case of Hideyo Noguchi, employed in the 1920s at the Rockefeller Institute, who infected hundreds of patients in New York's hospitals with syphilis for 'research purposes' (Corbellini and Lalli 2016). During the early 1900s several hundred people were also infected with syphilis and other sexually transmitted diseases in Guatemala. The 'research subjects' included orphan children. We know of several other studies conducted in the US that involved the injection of cancerous cells and exposure to radioactive substances, including uranium, and of many other studies in microbiology conducted similarly in China, Great Britain, Sweden, Italy and Russia.

Well after the end of the Second World War, and well after the Nuremberg Trials and the publication of the Nuremberg Code on ethical research, the Tuskegee Syphilis Study threw further discredit on science. In this study several hundred ill people from the Afro-American community were denied available medical treatment in order to observe and document the natural progression of the disease. Around the same time, it became known that Saul Krugman, employed by New York University, infected mentally ill children with hepatitis. Krugman gathered the parents' consent for, allegedly, vaccinations, but in reality children were fed with food contaminated with the faeces of ill patients, and were in that manner infected and 'studied'. In the 1970s it became known to the public that the CIA had performed a number of studies (under a programme called MKULTRA) which involved psychological torture and the use of various drugs, particularly LSD, with the aim of developing methods of mind control and mechanisms for coping with interrogation. It is unknown how many people were tortured and murdered, as the CIA has destroyed large parts of the evidence.

It should be noted that, with some exceptions, the 'results' of most of these studies have not become a part of the scientific literature (Corbellini and Lalli 2016: 92), because they were not based on any methodology. Thus, arguably, this was not science: these were sheer murders.

In any case, these examples provide a picture of incredible brutality and perversion, in which many thousands of people were victims over the course of the twentieth century (Corbellini and Lalli 2016: 92). And, of course, though it does not go so far, professional misconduct in science is of concern – in the last decade or so we have seen cases of the falsification of results, publication bias, and scientific and medical fraud. It can be argued, looking at human history, that exploitation and persecution (let alone fraud), in the name of science, religion or politics, are not isolated incidents – they are common. Usually these actions are directed at those perceived as 'others' (racial others, religious others, non-humans, and so on), not worthy of moral concern and respect. There is thus reason to be worried about those who have the power to exploit and persecute, and to vex, discriminate and abuse.

There is a further reason to be worried. We may think that the so-called Nazi scientists were psychopaths; that those other scientists involved in gratuitous torture and murder were also psychopaths, affected by the delirium of grandeur and moral viciousness. Actually, it is possible that they were all 'normal people'. We know from a number of studies in psychology performed and repeated since the 1960s that many ordinary people, not afflicted by any mental disorder, not morally callous and not psychopaths, can be turned into torturers or even potential murderers in the right (or rather wrong) circumstances (Milgram 1974; Zimbardo 2007).

What does this mean? It means that apprehension about science, which inherently gives people more control over others, and over the environment, is legitimate at some level. Scientists (or murderers in that disguise) have tortured and abused and murdered other humans over the course of history and have created tools that could lead to the destruction of entire cities in seconds. However, the apprehension is misplaced – it is not *science* or knowledge per se that should worry us. It is ourselves; it is human nature that should worry us. We possess an ability to dehumanise others that leads us, in certain contexts, to become brutal.

However, not all humans utilise this ability to become cruel. We may not naturally be gifted with the ability to recognise that the 'different' is equally valuable morally – and in the wrong cultural conditions, this can lead to the worst atrocities of which humans may be capable. It is indeed important to bear this in mind – that in certain social or cultural contexts, humans, not only collectively, but each of us individually, have the ability to become cruel. But the right cultural conditions may on the contrary enhance our empathy, enable us to recognise others as equally valuable, or at least to raise questions about who should be the subject of our moral concern and respect (Pinker 2012; Corbellini and Lalli 2016). In short, it is not science itself that perpetrates atrocities; it is humans, under certain cultural conditions. It is thus imperative that the right cultural conditions are established, and with this in mind Part II of this volume attempts to reason around regulatory mechanisms.

During the twentieth century, developments in biomedical sciences, including molecular biology and genetics, revolutionised the way human life is understood. Reanimation techniques raise the fundamental, metaphysical question of when it is that a person is dead (and thus, when it is that we are alive). New frontiers in reproduction, particularly the possibility of producing humans in alternative ways (with donors' gametes, or with 'parts' of gametes donated by third parties, or even with cloning techniques) raise metaphysical and legal questions relating to personal identity and parental rights. During the second half of the twentieth century it even became possible to create human DNA (recombinant DNA – rDNA). It is interesting to note that it was the scientists who themselves had created rDNA who asked for a moratorium on the development of science in this area, prior to elaborating guidelines for the continuation of research with the Asilomar Conference in 1975.

Scientific research requires complex negotiation, as we will see, *of values* between citizens, scientists, medical doctors, researchers, patients, research participants and society as a whole. Indeed, one could include non-humans in the pool of those whose interests should be considered in these negotiations. Political and scientific agendas may be at odds with one other. Many of the contributors here point this out. Political agendas are often inspired by the views of the majority in liberal democracies, or at least by the goal of finding viable compromises in areas in which views are starkly dialectical and dichotomous; and these two goals or aspirations may not be consistent with the aims and methods of scientific enquiry. Yet at a perhaps more profound level, politics and science should both be committed to the same ultimate goals: they should both serve people and society as a whole.

Thus, the questions posed in Part II of the collection are: how can politics better serve science? And how can science inform politics? The answers to these questions depend in part, of course, on the moral legitimacy and plausibility of specific scientific enterprises. Politics should not serve science if the purposes of science are malevolent, and of course politics should condemn moral turpitude in science, both in purposes and methods. But these answers also depend on other things: on how conflicts of interests may be resolved, for example; or on how accountability, through, for example, valid and reliable peer-review systems, may be achieved.

On this point, Ballabeni and Danovi (Chapter 9) highlight the pitfalls of the current peer-review system, established worldwide but rather outdated, they argue, and suggest alternative modes of adjudication regarding funding and the assessment of scientific validity; they also propose alternative modes of publishing structures in order to transform the communication of science to the general public, particularly by making use of the cheap and easily accessible World Wide Web. What seems to emerge from their analysis is that science and politics can reconcile their inherent tensions, but this requires the ambition to effect radical transformations to the cultural framework in which scientists operate.

Some of the authors in Part II argue that public or political recognition of the value of science raises specific political obligations. For example, Mertes (Chapter 11) suggests that the societal values that are lost through the tight regulation and prohibition of embryological research place an obligation on politicians to legislate, and legislate in one specific direction, the only one that is ethically defensible given the benefits and the losses at stake. Similarly, Cappato in Chapter 12 considers the tension between politics, regulation and science, in particular in the case of narcotic drugs. He argues that prohibition of the personal use of narcotic drugs, even if intended to protect citizens from the hideous consequences of addiction and to protect vulnerable members of society from exploitation (in the form of being caught in the net of illicit drug trafficking), still deprives society of important goods. Prohibition of narcotics has had as a by-product a comprehensive limitation on science, outlawing or heavily obstructing the medical use of illicit plants and substances, and research into their effects and potential. This approach has a number of consequences: one is that research on narcotics is inadvertently in this way 'handed over' to organised crime, which is more and more able to provide cheaper and more 'effective' (as well as more dangerous) recreational drugs. Second, it criminalises those in the grip of terminal or chronic illnesses, afflicted often by long-lasting and sometimes intractable pain, some of whom may find in opioids a valid form of pain control. Finally, prohibition results in a violation of fundamental human rights, Cappato argues. Scientific research and the enjoyment of its fruits are a human right: this human right is protected and defended by a number of UN declarations and conventions. To abide by the rights enshrined in the UN declarations and conventions, prohibitions should be radically reformed. Recent studies on the medical use of cannabis derivatives as well as some of the latest research on LSD and other controlled substances could be, if supported and promoted, a turning point in the matter.

Others highlight that there are other values at stake, not just the value of extending human life and ameliorating its quality. Baldoli and Radaelli (Chapter 14) explain how the precautionary principle is used to preserve and protect these other values, which can also be encompassed in the wider notion of societal benefits. Boggio and Romano (Chapter 10) discuss how freedom of research is codified in human rights law in the form of a human right to science, and articulate ways in which the right to science can be mobilised politically and judicially.

Scientific research, as has been pointed out earlier, is often perceived as a threat. However, Corbellini and Sirgiovanni (Chapter 13) point out that scientific research, and particularly scientific method, can actually protect us from this perceived threat. And by arguing this, they offer further considerations on how the tensions between politics and science can be resolved. They note that the prevailing theory about the relationship between science and human freedom is that science contributes to human autonomy or self-determination through the discovery of natural laws and by providing devices to solve practical problems in order to stimulate economic growth.

Simply put, the more we know about the world, the more we understand it, the better we can live in it, control it, and make autonomous, self-determined decisions about how we wish to conduct our lives. Countries in which science flourishes also tend to be middle- and high-income countries, and there is thus a positive association between economic growth and scientific freedom. But they note that a more likely hypothesis is that the invention and use of the scientific method in the modern age introduced into human communities a new way of thinking, which allowed a significant percentage of people to go beyond a set of cognitive and emotional biases that we inherited from our evolutionary ancestors, who, however, lived in simpler environments. In this way some psychological tools have been made available to an increasing number of people, prevalently in the Western world. These tools allowed human beings to achieve important cognitive and moral improvements, which made liberal and democratic governments possible: thus in a sense it is science that makes liberal democracies possible, and not liberal democracies that make (or should make) the progress of science possible. Science does not simply produce societal benefits that are tangible and usable (new vaccines, new forms of transplants); it also makes people more cooperative, less self-centred, less impulsive and more self-controlled (in the sense of autonomous), even in contexts that tend not to facilitate these behaviours.

One thing seems to follow from this: namely that politics should somehow recognise that it is science (in its many forms, and in its wider sense of the pursuit of knowledge through the systematic analysis of facts and reasoning around those) that generates and holds together, historically, psychologically and logically, the fabric of modern liberal democracies. Woolley (Chapter 15) points out that it is thus necessary to enhance the visibility of the research enterprise in society to ensure that decision-making by policymakers is responsive to scientific progress. Woolley calls on scientists to engage with non-scientists and actively advocate the value of research as a matter of public and national priority. When we talk about the right to science, or about freedom of scientific research, we should remember how the circulation and dissemination of scientific culture (including humanistic culture) can in itself promote the flourishing of society as a whole, and can even guarantee greater peace, in that it allows us to overcome certain moral and psychological boundaries (as Piccirillo notes in Chapter 8 of Part I) and recognise the equal value of others.

There are many specific areas of the regulation of science that this volume has left unexplored – genetics and genomics, research using non-human animals, biology (particularly synthetic biology) – and it will be interesting to evaluate how regulatory mechanisms will apply to artificial intelligence. Our aim is not to explore all areas of science that could give raise to ethical or political issues. Our hope is rather to stimulate reflection on important issues that affect many of us, and on the complexities inherent in the relationship between scientific research and regulatory mechanisms.

References

Corbellini, G., and Lalli, C. (2016), *Bioetica per perplessi*, Milan: Mondadori.

Frewer, S. A., and Schmidt, U. (eds) (2007), *History and Theory of Human Experimentation: The Declaration of Helsinki and Modern Medical Ethics*, Stuttgart: Franz Steiner Verlag.

Milgram, S. (1974), *Obedience to Authority: An Experimental View*, London: Tavistock.

Pinker, S. (2012), *The Better Angels of Our Nature: Why Violence Has Declined*, New York: Penguin.

Zimbardo, P. G. (2007), *The Lucifer Effect: Understanding How Good People Turn Evil*, New York: Random House.

9

Advocating a radical change in policies and new models to secure freedom and efficiency in funding and communication of science[1]

Andrea Ballabeni and Davide Danovi

A moving landscape

Threats and obstructions to scientific freedom, fairness and efficiency are commonly perceived as surrounding the scientific world. However, bottlenecks can also occur from within the system itself as some of the current regulations and forces shaping research (referred to here as 'science policies') substantially decrease the freedom and motivation of scientists. Indeed, inadequate policies can restrict the ability to perform research, the breadth of the fields of study, the methods of carrying out investigations and the dissemination of findings. Conversely, virtuous policies can have a positive impact on the freedom, quality of work and satisfaction of individual scientists. Importantly, from the societal perspective, the more scientists work freely, broadly and with motivation, the more they produce knowledge, data and innovation for the benefit of society, collectively ameliorating the scientific research system.

Unfortunately, 'science of science' is in its infancy. There has never been awareness of the importance of policies nor consistent interest in studying them, and specialised journals and articles are very few for several possible reasons. First, many researchers do not perceive the importance of policies for the functioning of science or may simply dislike 'science of science' studies and prefer to focus on their discipline. Second, many feel that their individual contribution would not be sufficient to affect the system and see no purpose in sacrificing their time and energy, as in the so-called 'paradox of voting' (Downs 1957: 25). Third, or a consequence of all the above, with few academic positions and funding opportunities, the field is at the moment not profitable with regard to career returns.

Nonetheless in order to promote research and discussion on policies with a view to improving them, it is of great importance that science policy experts reach out to the rest of the scientific community. It is key to spread awareness of the many possible changes (however small or big) that can affect the freedom, fairness and efficiency of scientific work. Moreover, it is crucial that public scientific agencies incentivise investigations by mobilising funding and nurturing opportunities. In this regard, it is advisable that scholars approach this area, combining a thorough and systematic analysis of the current situation with new perspectives including radical and 'out-of-the-box' approaches. There are several aspects of the process of sourcing, obtaining and sharing scientific research that can be improved by implementing effective policies. In this chapter, we focus on two main areas: funding and communication. We discuss the current situation, highlighting what we believe are the problems and the ongoing efforts to find solutions, and then moving to propose paradigm shifts that, albeit radical, appear to us as the natural progression of the existing trends. We concentrate on research in life sciences, our area of expertise, proposing that the matters treated can extend to broader areas of investigation.

Funding of science: ideas for an alternative system

Scientific research, especially when it requires costly equipment as most modern biomedical research does, is an expensive endeavour and good funding is key. There is a very broad consensus over the fact that research is currently underfunded all over the world. The relationship between the funding of specific research fields and their deliverables is a matter of constant debate. Basic research in particular is resistant to evaluation in terms of impact (and thus return on investment), intrinsically requiring long timescales and a dose of serendipity, and often incompatible with the current system of resources and careers. In fact, human beings are consciously and unconsciously biased towards valuing the present more than the future, and inherently show preference for certain outcomes over uncertain risks. Furthermore, the allocation of funding is usually under the direct or indirect influence of politicians, who favour rapid and visible returns over long-term strategies due to short-term electoral cycles. Altogether, these factors have a profound influence on the availability of resources and the efficiency of scientific research. While maintaining the need for increasing investment in scientific research overall, we focus in this section first on describing the current system and subsequently on proposing changes in funding distribution.

At the present time, scientists receive monies to perform their research largely through the submission of proposals to governmental or private funding bodies. These recruit panels of other scientists, usually indicated as 'peers', to evaluate whether the proposed projects are worth consideration. To prepare a proposal, a considerable amount of time is spent in writing and on associated paperwork. After submission, the peer-review process can take several more months. At the end of this process, if the project is approved, a certain amount of money is allocated to that specific project. On the contrary, if the funding is not approved, that project does not receive any money. An increasing trend in Western countries is that most proposals are not funded. For example, the rate of success for grant proposals submitted to the NIH (the National Institute of Health in the US) is less than 20 per cent (Alberts et al. 2014). The total grant success rate in the MRC (Medical Research Council, UK) is 23 per cent (MRC 2014–15). Submitting more than a few proposals per year is generally prohibitive for a principal investigator. Given the limited number of opportunities and the time and energy costs, missing even two funding attempts in a row could have significant impact on a research group. Indeed in such cases investigators, particularly younger ones, often have to quit their research and disassemble their team.

The acceptance of this model by the scientific community appears to us surprising. In our experience, voicing discontent beyond private conversations is perceived as an implicit admission of failure and the poor quality of research, when coming from young researchers, or an unnecessary attempt to obtain extra publicity when coming from established ones. Unfortunately, this 'all-or-nothing' scheme does not do justice to the gradual distribution of merits and skills among scientists. In addition, the prediction of success is elusive, as in most cases the research plan can differ substantially from that proposed, as milestones and deliverables, where present, can be reformulated. Furthermore, the edge that established scientists have over younger ones is well known under the name of the 'incumbency advantage' (Ballabeni et al. 2016). This phenomenon has several different causes. The complexity of the procedures for grant applications might be a handicap for newly established scientists who are less experienced or cannot afford assistants (Daniels 2015). Grant applications are commonly centred on the assessment of a significant amount of preliminary data, which obviously favours established investigators and bigger laboratories (Alberts et al. 2014; McDowell et al. 2014; Daniels 2015). Moreover, there is evidence of bias in favour of the 'insiders and the familiar' over the 'unknown' (Daniels 2015; Fang and Casadevall 2009; Kirwan Institute 2014; Nicholson and Ioannidis 2012) and against unconventional and unorthodox ideas, which are more likely to be proposed by newly established scientists. The latter are also less likely to be involved in large research programmes, which can be favoured over small ones by 'inertia' and 'financial dependency' biases (Alberts et al. 2014). Finally, the evaluation of proposals usually takes into consideration the record of publications, which is itself affected by 'incumbency advantage' (see below). In essence this scenario not only entails reduced freedom of research and significantly sustains inequality in the system, but it also instils stagnation of ideas. Importantly, it produces 'overfunded' (i.e. inefficiently funded) laboratories that experience diseconomies of scale (Berg 2010; Stephan 2012; Alberts et al. 2014; Woolston 2015) and diverts funding from innovative projects, forcing many junior scientists to quit research. The dire consequences of this are difficult to fully capture for the community and society.

Some routes to alleviate these problems are being explored. The changing scenery includes examples of crowdfunding (e.g. Cancer Research UK 2014; Vachelard et al. 2016; Crowd Science 2017; Crowdhelix n.d.) and numerous public–private ventures that coexist as non-canonical ways to support research. In parallel, interesting new tools are being developed that use big data mining to allow more transparent access to the research grants and fellowships awarded to laboratories (e.g. Symplectics Research Information Management System). However, the vast majority of funding for basic research continues to be assigned through proposals submitted to funding agencies and evaluated by the mechanism of the 'peer review'. To tackle the incumbency advantage of this system a few ideas have been tested or proposed. One is capping the amount of funding to a research group with a view to funding more scientists, which has been thoroughly discussed within the scientific community (Alberts et al. 2014; McDowell 2014; Ballabeni et al. 2016). Furthermore, a small number of agencies have been assigning funds by evaluating the overall quality of scientists rather than the research project (Alberts et al. 2014; Daniels 2015; Ballabeni et al. 2016). Importantly, the latter change tackles the problem of the long and time-consuming research proposal preparations and the tendency to disproportionately prefer incremental mainstream projects over more risky and radical ones. Data from a recent broad survey (Scita et al. 2016) show that a consistent majority of life scientists, regardless of their career stage, are in favour of these policies. Assigning funding based on general reputation rather than on project proposals might appear to increase the detested 'incumbency advantage'. On the contrary, we believe it is likely, as a recent survey indicated (Ballabeni et al. 2016), that promoting caps on funding and a more gradual allocation of funding would be effective in reducing the 'incumbency advantage'. Another criticism comes from the fact that salesmanship may be favoured, although arguably no more than in the existing grants scenario.

These ideas for new policies might therefore significantly increase fairness by giving everyone similar initial opportunities and by judging the relative merits of scientists through a more proportional lens. In general, the advantages of having an additional number of rewarded scientists and a more widespread distribution of resources are perceived as greater than the disadvantages of decreasing the average funding per scientist. In addition, providing funding to the scientists rather than to the projects would increase efficiency by cutting the time and energy spent on proposal writing and by supporting riskier and potentially ground-breaking research. Additionally, these transformations might also increase the efficiency of the system, as setting caps for investigators or research groups would increase the pool of funded scientists, especially among the younger ones. And by keeping young researchers in the system, it would develop a well-structured workforce for the next decades.

However, is it possible to imagine an even more radical change, a leap into a totally different system? A peer-review system based not on panels of scientists but on collective wisdom? (Bollen 2014; Ballabeni et al. 2016). This 'lateral thinking' idea is very recent and not yet common knowledge in the scientific community. It is based on a highly decentralised system in which the wisdom of the whole scientific community is used to allocate funding by avoiding the current 'all-or-nothing' approach. Here, being a researcher (for which minimal criteria should be satisfied, for example, a specific certified degree and/or some form of entry-level check) gives one the right to be considered equally with all other researchers under the same funding initiative. The funding agency would give all scientists under its competence an equal amount of funds per year. However, scientists would be required to allocate a fixed percentage of their last year's funding to colleagues whom they think deserve it. As a consequence, each year researchers would receive a fixed basic amount of money from a funding agency plus an elective contribution decided and donated by their colleagues. While still requiring human judgement, this system would nonetheless cut substantially the huge operating costs of panel-based peer review, and might also neutralise the biases and inefficiencies of the current system. Clearly, the assumption is that most scientists would allocate funding on a purely meritocratic basis, with the greatest interest in maximising fairness, the advancement of knowledge and the improvement of society. Basically, those scientists who are anticipated to lead the best science would accumulate more. On the other hand, scientists who receive more must in turn distribute a larger amount of money; their prominence would therefore give them greater influence on how resources are used. And the constant yearly basic grant would ensure more stability and provide greater autonomy, while it still might be limiting in isolation, encouraging meritocratic advancements. In addition, as the preferences of the scientific community evolve, so the pattern of funding would adapt according to the wisdom of the scientific crowd, giving the possibility of fine-tuning funding levels to where they are most needed.

There are some foreseeable criticisms of this idea. It might at times be difficult to claim that one defined research outcome has been funded by one funding body in particular, so this system might not apply to charities or foundations. Moreover, anonymity, a key element in this system, would need to be enforced to avoid illegitimate practices such as, for example, the creation of circular schemes. However, one could argue that concerns about circular schemes have already been voiced in the current system (e.g. Ghosh 2010). In addition, assigning funding based on general reputation rather than on project proposals might appear to increase the 'incumbency advantage'. On the other hand, it is likely that promoting a more gradual allocation of funding would be effective in reducing this. Indeed, contrary to the panel-based peer reviews based on panels, which are often composed of established scholars, in this more horizontal system junior investigators would have a bigger role in deciding how to allocate funding, and thus might plausibly be more likely to know and credit their peers. Furthermore, despite the amount of paperwork involved, some scientists value writing grant proposals as a way of 'crystallising' thoughts into a defined plan of action. Yet we contend that the milestones and deliverables of a project could still be present, and would probably better mirror the reality of scientific projects than in the current system.

In any case, we believe that the benefits outweigh the possible drawbacks. Indeed, this system would not only decrease the financial and time costs of panel-based peer review, but it would also cut the time taken up by writing grant proposals, as it is conceivable that such proposals would be less and less important; indeed, much more emphasis would probably be placed on the overall evaluation of a scientist's scientific and delivery skills. What's more, it would not involve an 'all-or-nothing' allocation of funding and this would guarantee a more proportional, and thus more meritocratic, distribution of resources. Likewise, this system would assure a safety buffer for difficult times. It would dynamically and finely adjust the allocation of funding according to the community interests. Additionally, it would be based on the potential evaluation of more people than those present in a panel, guaranteeing a broader, more unbiased and continuous examination. Finally, it might promote profound changes in scientific communication, as scientists will more openly share past achievements, current data and future plans.

In conclusion, we see the mindset around the funding of science as beginning to change. New and ongoing paradigm-shifting initiatives, like the ones we describe, will lead to a potentially complementary, more functioning and beneficial research system.

Communication of science: ideas for an alternative system

Similar to and possibly more so than in the funding arena, the way science is communicated is constantly debated. At its core, the current system of science publication is over 350 years old. Scientists present the results of their research in the form of scientific articles (papers) sent to journals that are owned and managed by publishing corporations. The format is generally locked into the undoubtedly beneficial logic of an introduction–methods–results–discussion stream. Upon receiving a paper, the editors of a journal can reject it immediately or send it to peer reviewers (also known as referees). These are generally anonymous to the authors of the papers, and the process is therefore referred to as single-blind peer review. The editors use the peer review produced by referees to decide between the following options: immediate rejection, requests for improvement or (anecdotally) immediate acceptance of the article. The entire process normally takes several months (or sometimes years) from the first submission, and an article can frequently go through multiple rounds of peer review and revision. After a paper is accepted it is published sometimes weeks or months later, and then it usually cannot be peer reviewed, corrected or generally even commented on any more. Only in rare circumstances are mistakes corrected and only very rarely is a paper retracted if major flaws or actual frauds are proven. Papers help the dissemination of findings and ideas while advancing the scientists' careers. More papers being published in the most read journals translates into more opportunities for funding and promotion. This system has for decades served scientific research while granting to successful publishing corporations the luxury of selecting a very small fraction of scientific output from authors, simultaneously sourcing the scientific output of a pool of reviewers at no cost. Thus, especially in recent times an increasing number of problems have arisen around both the reviewing process and access to publication.

As a matter of fact, peer reviewing can be at times a pretty sloppy process. Anonymity and lack of incentives anchor the acceptance of the role and the quality of a referee's work to their awareness of prepublished science and loyalty to a greater good, respectively. In a minority of cases, because of rivalry, sympathy or dislike, unfair peer reviews are deliberately provided. Also, it does happen sometimes that the referee privately reveals her or his identity to the authors, and in fact some journals compel or allow reviewers not to be anonymous.

Whereas scientific progress is dynamic, by definition the publication system is static. Papers cannot generally be improved once they are published. In some cases, editors can offer the possibility of publishing corrections. Continuous improvement to create new versions (a process also referred to as 'versioning', which is extremely common for software and scripts) does not exist except for very few exceptions (e.g. Singh 2015). In very rare circumstances, if the data are proven to be simply wrong or if fraud is demonstrated, an article can be retracted. However, it is now increasingly evident that fraudulent behaviour is spotted and signalled very rarely, and that the very small number of retracted papers represents only a fraction of *de facto* unreproducible results (Baker 2016). This is because neither editors nor reviewers nor (obviously) authors have any interest in spotting flaws after a paper is published. In essence, there is no further peer review after publication. So, the paper is first peer reviewed by just two to four reviewers before publication, whereas afterwards it does not receive any further similar peer review. And without the possibility of post-publication peer review and open commenting on papers, flaws are even less likely to be spotted and signalled. This is unfortunate considering that hundreds of (not anonymous) scientists might be willing to provide their views. Open commenting on papers (comments from scientists or laypeople, similar to what happens in blogs) happens extremely rarely.

The consequences of this publication landscape are perceivable in the fake sense of the official quality and validity of publications, which scientists have learned to take for granted when they are not able to reproduce results from a colleague's publication. Moreover, many readers, and less so the general public, tend to perceive published articles positively only because these articles are in journals, especially journals with a high impact factor. Scientists who want to stretch or overemphasise their data only have to pass the test of a few people before their message enters a comfort zone in which it will probably never be scrutinised directly again.

Furthermore, there is a predominant culture in academia of novelty over thoroughness and of creativity over robustness. There are fashions and trends in scientific fields that influence perceived importance (Pfeiffer and Hoffmann 2009). There is a diehard snobbery regarding 'negative results' (Goodchild van Hilten 2015), even in clinical trials where debate is raging (AllTrials 2015). In addition, the constrained format pushes some research out of the 'publishable' area. Our brains are wired to the narrative of a 'story' and this makes it more difficult for entire fields of research (research involving different types of 'screening' is a good example) to be considered 'scientific' and not merely 'technical'. Furthermore, the economic model of publishing prescribes a return for the shareholders, and sustains very expensive subscription fees or tariffs for single-article access. This is obviously a major problem for the scientific community as accessing scientific literature can drain a lot of financial resources and fuel the 'incumbency advantage'. It is also unfair for the general public as it precludes access to research, including in some cases research funded via taxation.

Shockingly to us, limiting the number of published papers, irrespective of quality, is a policy choice of established journals even when the actual throughput could be much higher. The strengths of a scientific study in terms of novelty, elegance, reproducibility and impact have in some respect more to do with art than science, and *where* (i.e. in which journal) often matters as much as *what* (i.e. what study) is published. Can the queue outside an exclusive club be a better measure of success than the number of people on the dance floor or their ability?

We contend that the internet has pushed forward the revolutionary change of publishing 'virtually' as opposed to 'on paper', at a fraction of the cost. Scientists can therefore opt to disseminate research in a more open, flexible and collective way (and are beginning to do so). In the last few years more and more articles have been published that are freely accessible to anyone, either in fully 'open access' journals or in hybrid models. Yet surprisingly, and even for small specialised journals, the majority of newly published articles are still 'closed access' (Lewis 2012). Beside 'open access', another change that is driven by the internet is the so-called 'preprint publication'. These are articles published by journals or on websites (e.g. ASAPbio, bioRxiv, PeerJ, F1000Research, Figshare) that do not require any previous peer review. Yet the vast majority of data that makes it to the scientific community is hampered by the long lead times (usually several months to several years) required to complete the submission and peer-review process, and in many cases still has restricted access.

We believe that open access and the avoidance of peer review before publication, as in preprints, is having an impact. Yet offering flexibility in formats, the possibility of post-publication peer review, versioning and incentives for peer reviewers could lead to even more radical strategies at the system level that should be trialled and could make a real difference. A key element of the system we propose is a central international repository held as a reference for each scientist, to enable evaluation for funding, hiring and promotion by assessing all their activities. Here, peer review and original research publications would have complementary value and legitimacy and this will naturally incentivise the reviewing process. The technical, financial and political capacity to establish and maintain such an infrastructure is already within reach. In this system scientists would publish their data and ideas freely, without restrictions of style and format, choosing whether to have peer review before publication or to publish without any previous peer review. However, publications could be effectively open to peer review after publication. The people who perform this type of peer review, also called 'post-publication peer review', could be investigators proposed by the authors, investigators proposed by evaluators of the article or the authors (hiring panels, funding agencies, prize committees etc.); or, simply investigators who are interested in providing their peer review. Peer review will, then, be fully transparent. The reviewers would be identified and their comments made public. The reviewers, being fully accountable, would have a genuine incentive to produce fair and accurate comments and recommendations. Articles could be easily modified through 'versioning'. Each version (including the original) would have to remain online with a unique date of publication and digital object identifier (DOI). Each version would be open to peer review and possibly also comments from the general public (similarly to what happens in blogs and social media). In this scheme it is conceivable that once an article is published, it would not have the possibility of being retracted. Scientists would modify their articles through versioning, with each version open to the scrutiny of the scientific community, rendering actual retraction redundant.

At first these proposals might look utopian or naive to scholars who have grown up in the current system. We believe it would not be so in a mutated cultural framework and that the scientific community should have the courage and ambition to offer radically new opportunities to the present system that could serve science and society. Admittedly, what we suggest presents some problems too. It is, for example, possible that some publications could simply present preliminary data. Moreover, the increase in format and design options would make publications more difficult to read, especially in the beginning. An additional potential problem could be the possible deluge of information that would be poured on to the internet as a result of the lack of prepublication gates. However, we believe that in the long run continuous scrutiny and versioning as well as the incentive of being an 'active reviewer' would push forward more accurate, complete and effective science communication paradigms. The 'authority' of an 'active review' may also be weighed against its author's previous work (as proposed in Brewer 2014). Our perception is that many low-ranking journals (with poor or, even worse, arranged peer reviews) are already flooding the internet and the proposed change would not have such a dramatic impact on the total number of publications. The best research studies will naturally emerge with profiles of views and citations over time. We contend that the advantages would be far greater than the disadvantages, especially in the long run. We obviously do not propose that this system should abruptly replace journals that have served well for decades, and that may well continue in parallel with a complementary role.

This radically different publishing structure, based on the full use of the internet's capabilities, could change the system of incentives for scientists and reviewers and would tackle the problems listed above associated with the contemporary system born in a pre-internet and pre-computer era. It would provide much more freedom in format and design, easing scientists' work and guaranteeing that more data would be published and disseminated, significantly decreasing publication time. Also, as peer reviewers would be identified and evaluated on their peer-review work as much as for their primary research work, the current sloppiness of the peer-review process would be avoided or significantly decreased. Thus, unfair peer reviews would be prevented as well as the private disclosure of the identity of referees to authors. Moreover, peer review and open commenting for all articles would be incentivised after publication, increasing both scientific discussion and scrutiny. In addition, there would be the possibility of producing improved versions of an article. As each version would remain online, authors would still be discouraged from publishing wrong or poor information. Shoddy or fraudulent behaviour would be policed by the possibility of open post-publication scrutiny. Furthermore, contrary to the current system, the publication in itself would not convey any fake sense of validity or quality to an article. In this new system the publication would have less value and the focus would instead be shifted on to the content of the article and the peer review. Finally, this truly 'open access' system would be accessible to lay persons and would allow scientific institutions to save a lot of money that could be used to advance science more effectively. This is especially deserved when taxpayers have funded the research.

In this chapter we have proposed two radical system-level changes for the funding and communication of scientific research. Although other, alternative, complementary, small or big changes could also be possible, we believe that these two propositions would have a significant impact on scientists and society. To build momentum it would first be necessary to raise awareness about the current problems and the feasibility of a radically different system through more public debate, online as well as in meetings. As awareness raising by itself will not be sufficient, it will be also crucial to promote more policy-focused research and discussion. Understanding the motivations, goals and views of scientists will be key to designing policies that can increase freedom and motivation as well as maximising the advancement of knowledge and practical benefits to society (Ballabeni et al. 2014; Scita et al. 2016).

In conclusion, we believe that these paradigm-shifting proposals are an example of how the freedom and quality of work of scientists as well as the fairness and efficiency of the research system can be significantly improved by designing far-reaching and ground-breaking policies.

Note

1 We wish to thank our colleagues for scientific discussions, our funders and our host institutions. We are writing here in a personal capacity and our views are not necessarily shared by them.

References

Alberts, B., Kirschner, M. W., and Tilghman, S. et al. (2014), 'Rescuing US biomedical research from its systemic flaws', *Proceedings of the National Academy of Sciences of the United States of America*, 111.16: 5773–7.

AllTrials (2015), 'Half of all clinical trials have never reported results', www.alltrials.net/news/half-of-all-trials-unreported (last accessed 27 October 2017).

Baker, M. (2016), 'Biotech giant publishes failures to confirm high-profile science', *Nature*, 530.7589: 141.

Ballabeni, A., Boggio, A., and Hemenway, D. (2014), 'Policies to increase the social value of science and the scientist satisfaction: an exploratory survey among Harvard bioscientists', *F1000Research*, 3: 20.

Ballabeni, A., Hemenway, D., and Scita, G. (2016), 'Time to tackle the incumbency advantage in science: a survey of scientists shows strong support for funding policies that would distribute funds more evenly among laboratories and thereby benefit new and smaller research groups', *EMBO Reports*, 17.9: 1254–6.

Berg, J. (2010), 'Another look at measuring the scientific output and impact of NIGMS grants', http://go.nature.com/dae21z (last accessed 27 October 2017).

Bollen, J. Crandall, D., and Junk, D. et al. (2014), 'From funding agencies to scientific agency: collective allocation of science funding as an alternative to peer review', *EMBO Reports*, 15.2: 131–3.

Brewer, J. H. (2014), 'o'Peer: open peer review', *Journal of Physics: Conference Series*, 551: 1, article ID. 012060.

Cancer Research UK (2014), 'Crowd funding cancer research and how you can help', http://scienceblog.cancerresearchuk.org/2014/11/14/crowdfunding-cancer-research-our-latest-experiment-and-how-you-can-help (last accessed 27 October 2017).

Crowdhelix (n.d.), www.crowdhelix.com (last accessed 3 November 2017).

Daniels, R. J. (2015), 'A generation at risk: young investigators and the future of the biomedical workforce', *Proceedings of the National Academy of Sciences of the United States of America*, 112.2: 313–18.

Downs, A. (1957), *An Economic Theory of Democracy*, New York: Harper & Row.

Fang, F. C., and Casadevall, A. (2009), 'NIH peer review reform – change we need, or lipstick on a pig?', *Infection and Immunity*, 77.3: 929–32.

Ghosh, P. (2010), 'Journal stem cell work blocked', *BBC News*, 2 February, http://news.bbc.co.uk/2/hi/8490291.stm (last accessed 27 October 2017).

Goodchild van Hilten, L. (2015), 'Why it's time to publish research "failures". Publishing bias favors positive results; now there's a movement to change that', Elsevier, https://www.elsevier.com/connect/scientists-we-want-your-negative-results-too (last accessed 27 October 2017).

Kirwan Institute (2014), *State of the Science: Implicit Bias Review*, http://kirwaninstitute.osu.edu/wp-content/uploads/2014/03/2014-implicit-bias.pdf (last accessed 27 October 2017).

Lewis, D. W. (2012), 'The inevitability of open access', *College & Research Libraries*, 75.5: 493–506.

McDowell, G. S., Gunsalus, K. T. W., and MacKellar, D. C. et al. (2014), 'Shaping the future of research: a perspective from junior scientists', *F1000Research*, 3: 291.

MRC (2014–15), www.mrc.ac.uk/research/funded-research/success-rates/all-published-success-rate-data/grant-success-rates-2014-151 (last accessed 27 October 2017).

Nicholson, J. M., and Ioannidis, J. P. (2012), 'Research grants: conform and be funded', *Nature*, 492.7427: 34–6.

Pfeiffer, T., and Hoffmann, R. (2009), 'Large-acale assessment of the effect of popularity on the reliability of research', *PLoS ONE*, 4.6: e5996, http://journals.plos.org/plosone/article?id=10.1371/journal.pone.0005996 (last accessed 27 October 2017).

Scita, G., Sorrentino, C., and Boggio, A. et al. (2016), 'Increasing the public health potential of basic research and the scientist satisfaction: an international survey of bioscientists', *F1000Research*, 5: 56.

Singh, C. D. (2015), '"Living figures" make their debut', *Nature*, 521.7550: 112, www.nature.com/news/living-figures-make-their-debut-1.17382 (last accessed 27 October 2017).

Stephan, P. (2012), 'Research efficiency: perverse incentives', *Nature*, 484.7392: 29–31.

Woolston, C. (2015), 'Bigger is not better when it comes to lab size', *Nature*, 518.7538: 141.

Vachelard, J., Gambarra-Soares, T., and Augustini, G. et al. (2016), 'A guide to scientific crowdfunding', http://journals.plos.org/plosbiology/article?id=10.1371/journal.pbio.1002373 (last accessed 27 October 2017).

10

Freedom of research and the right to science: from theory to advocacy

Andrea Boggio and Cesare P. R. Romano

Although the right to science, which includes both the right of scientists to do research and the right of everyone to benefit from that research, was recognised internationally as early as 1948, it is arguably the least known, discussed and enforced international human right. As a result, its binding normative content is not settled and needs to be better clarified and specified. Progress at the conceptual level has been made in the last few years but we are still far from a full understanding of this right and its normative content, and from having a cohesive and authoritative list of duties that states must abide by to fully realise the right.

In this chapter, we argue that legal and political mobilisation in international forums provides promising paths to further define the normative content of the right to science. Waiting for the theoretical debate on the right to science to settle before seeking its protection would delay its realisation. Mobilisation through advocacy and litigation can provide both a remedy to victims of violations in specific cases and cause the development of a body of opinions and other policy outcomes which can contribute, with authority, to defining the content of the right.

In the first part of this chapter, we map out the recognition of the right to science under international law, both at the global and regional level. We then look at important international developments, and in particular, the work of the United Nations Special Rapporteur on Cultural Rights, and the emergence of an academic debate on the right to science. We then turn to legal and political strategies to mobilise the right to science. By 'legal mobilisation' we mean the use of courts and tribunals (i.e. judicial remedies) to seek vindication of the right to science for violation of this right. We identify international judicial and quasi-judicial institutions that have jurisdiction over violations of this right, and discuss the procedural requirements and some of the challenges claimants face. With regard to 'political mobilisation', we identify venues where human rights advocates, scientific societies and other civil society organisations could push for the realisation of the right to science. At the global level, we identify opportunities for political mobilisation in connection with the United Nations Human Rights Council's Universal Periodic Review, the State Reporting Procedures and the Special Mandates. Finally, we identify opportunities at the regional level.

Legal recognition of the right to science

The right to science is as old as international human rights. It was recognised first in 1948, in the Universal Declaration of Human Rights, the keystone of the international human rights architecture. Article 27 of the Universal Declaration of Human Rights provides that:

(1) Everyone has the right freely to participate in the cultural life of the community, to enjoy the arts and to share in scientific advancement and its benefits.
(2) Everyone has the right to the protection of the moral and material interests resulting from any scientific, literary or artistic production of which he is the author.

This right found further recognition in 1966, in the International Covenant on Economic, Social and Cultural Rights (ICESCR), a multilateral treaty adopted by the United Nations General Assembly (in force from 3 January 1976). Under Article 15:

1. The States Parties to the present Covenant recognize the right of everyone:
   (a) To take part in cultural life;
   (b) To enjoy the benefits of scientific progress and its applications;
   (c) To benefit from the protection of the moral and material interests resulting from any scientific, literary or artistic production of which he is the author.
2. The steps to be taken by the States Parties to the present Covenant to achieve the full realization of this right shall include those necessary for the conservation, the development and the diffusion of science and culture.
3. The States Parties to the present Covenant undertake to respect the freedom indispensable for scientific research and creative activity.
4. The States Parties to the present Covenant recognize the benefits to be derived from the encouragement and development of international contacts and co-operation in the scientific and cultural fields.

The combined reading of these two provisions provides the legal foundations of what is now commonly referred to as the 'right to science' (Besson 2015: 404), or, less succinctly, the 'right to enjoy the benefits of scientific and technological progress and its applications'.

Various legal instruments at the regional level also recognise the right to science. In Europe, there is no reference to the right to science either in the European Convention on Human Rights (1950) or in the European

Social Charter (1961; revised 1996), two of the most important human rights treaties in Europe. However, at least as concerns the European Union, this lacuna was filled in 2000 with the adoption of the Charter of Fundamental Rights of the European Union, which provides that scientific research shall be 'free of constraint'.

In the Americas, one can find several relevant provisions in the Charter of the Organization of American States (1948), the most relevant being Articles 17, 30, 34.i, 38, 45, 47 and 51. It is also mentioned in Article XIII of the American Declaration of the Rights and Duties of Man (1948, which provides that '[e]very person has the right . . . to participate in the benefits which result from intellectual progress, especially scientific discoveries'. Finally, it is mentioned in the Additional Protocol to the American Convention on Human Rights in the Area of Economic, Social and Cultural Rights ('Protocol of San Salvador', 1988), which requires states to recognise the right of everyone 'to enjoy the benefits of scientific and technological progress' (art. 14.1.b) and 'extend among themselves the benefits of science and technology by encouraging the exchange and utilisation of scientific and technological knowledge' (art. 38).

In Africa, the Charter of the African Union (1963) identifies scientific and technical cooperation as essential for meeting its goals (art. II (2)), and the Protocol on the Rights of Women in Africa of the African Charter on Human and Peoples' Rights (2003) requires states to take specific measures to promote education and training for women, particularly in the fields of science and technology (art. 12 (2)(b)).

In the Arab world, the Arab Charter on Human Rights (2004) recognises the right of everyone 'to take part in cultural life and to enjoy the benefits of scientific progress and its application', together with the obligations of states to

> respect the freedom of scientific research and creative activity . . . ensure the
> protection of moral and material interests resulting from scientific, literary
> and artistic production . . . enhance cooperation at all levels, with the full
> participation of intellectuals and inventors and their organisations, in order
> to develop and implement recreational, cultural, artistic and scientific pro-
> grammes. (art. 42)

Finally, in South East Asia, the ASEAN Human Rights Declaration (2012) provides that every person has the right, individually or in association with others, to freely take part in cultural life, to enjoy the arts and the benefits of scientific progress and its applications and to benefit from the protection of the moral and material interests resulting from any scientific, literary or appropriate artistic production of which one is the author (art. 32).

International initiatives

Although the right to science has been recognised under international law since 1948, international, regional and national bodies, as well as human rights activists and scholars, have paid little attention to it. As a result, our understanding of the normative content of the right to science – that is, what exactly are states' obligations – is not entirely settled. However, in the past two decades the right to science has moved to the front and centre of the debate in international forums, and progress towards a more complete understanding of this right has been tangible. Two developments, at the global level, are particularly significant: first the adoption, under the auspices of UNESCO, of the Venice Statement on the Right to Enjoy the Benefits of Scientific Progress and its Applications ('Venice Statement', 2009); and, second, the appointment by the Human Rights Council of a Special Rapporteur in the field of Cultural Rights, whose mandate also includes the right to science.

The Venice Statement was the outcome of a 2009 meeting sponsored by UNESCO aiming at 'clarifying the normative content of the right to enjoy the benefits of scientific progress and its applications and generating a discussion among all relevant stakeholders with a view to enhance the implementation of this right'. The Venice Statement makes two significant contributions. The first is to spell out the three duties which states parties to the ICESCR have: the duty to respect, to protect and to fulfil. 'Respecting' means guaranteeing the freedoms which are necessary to do science (e.g. autonomy, freedom of speech, freedom to assemble in professional societies and to collaborate). 'Protecting' means ensuring that science is not done by infringing upon the rights of anybody (e.g. research subjects, vulnerable populations). 'Fulfilling' calls for a variety of strategies including monitoring harms arising from science, enhancing public engagement in decision-making about science and technology, ensuring access to the benefits of scientific progress on a non-discriminatory basis, and developing science curricula at all levels of schooling. Second, the Statement points out that it is also incumbent upon non-governmental actors (e.g. scientific societies, for-profit entities, civil society) to contribute to the realisation of the right to science. The Statement touches upon the issue of the privatisation of science and how it could conflict with the right to science.

The second significant development at the global level is the United Nations Human Rights Council's decision to give a Special Rapporteur a mandate on cultural rights, including the right to science (Resolution 10/23). The first appointee was the Pakistani sociologist Farida Shaheed, and the current one is the Algerian-American law professor Karima Bennoune. In 2011, Farida Shaheed visited several UN members and organised a public consultation in Geneva under the auspices of the Office of the United Nations High Commissioner for Human Rights. Member states' civil society organisations were asked to fill out a questionnaire that the Special Rapporteur later used in her report, entitled 'The right to enjoy the benefits of scientific progress and its applications', released in 2012 (United Nations 2012).

This report is a fundamental contribution to the field as it discusses the right to science from different angles: its normative content, state obligations and its limitations. With regard to the normative content, the report makes four contributions. First, it connects the right to science to the right to participate freely in the cultural life of the community as recognised by Article 15 of the ICESCR. Article 15 entails the right to contribute to science (as knowledge producers) and enjoy opportunities to participate in decisions about science (as citizens). The report further maintains that the right should be enjoyed free of discrimination. Second, it stresses the importance of freedom of research as a prerequisite of the enjoyment of the right to science. In fact, the ability to 'continuously engage in critical thinking about themselves and the world they inhabit, and . . . the opportunity and wherewithal to interrogate, investigate and contribute new knowledge with ideas, expressions and innovative applications, regardless of frontiers' are prerequisites for implementing both rights (para. 18). Third, it connects the right to science to the concept of human dignity to the extent that the right protects people's 'ability to aspire – namely, to conceive of a better future that is not only desirable but attainable' (para. 20). Aspirations, the Rapporteur notes, 'embody people's conceptions of elements deemed essential for a life with dignity' (para. 20). Fourth, it identifies links to other rights. In some cases, the right to science is enjoyed in conjunction with other rights, such as the right to seek information, to take part in the conduct of public affairs, to self-determination, to development, and to make informed decisions (paras 21–2). The right to science is also a prerequisite for the realisation of other rights, namely the right to food, health, water, housing, education and a clean and healthy environment (para. 23).

The second part of the report focuses on the normative content and related obligations of states. In this section, the Rapporteur proposes a list of objectives which states must guarantee: access by all without discrimination; freedom of scientific research and opportunities for all to contribute to the scientific enterprise; individual and collective participation in decision-making; and an environment which enables knowledge production and exchange.

The last section of the report discusses the limitations of the right to science. The Rapporteur points out that limitations certainly arise from the very same body of human rights law and, thus, it must promote general welfare and be proportionate to the objective (para. 49). The regulation of research subjects provides an example of a justifiable limitation of the right to science (para. 51). More controversially, the Rapporteur also cites the precautionary principle as an important guide for science and technology policies in the absence of scientific consent such that a certain sense of caution would not cause irreparable harm to the public or the environment (para. 50).[1]

Academic debate

The academic debate has mainly taken place between a handful of scholars whose work primarily focuses on refining the theoretical framework for thinking about this right and defying its place in human rights law. Schabas (2007) argues that states must respect scientists' freedom to conduct research, build facilities for research, preserve minorities' cultural rights and protect the rights of indigenous peoples. Chapman (2009) identifies three rights: to access the benefits of scientific progress and technology without being discriminated against; to be protected from the harmful effects of science and technology; and to protect individuals' intellectual property. Muller (2010) argues that states must create 'an institutional framework and [adopt] policies and laws in relation to science and technology that enable individuals to freely conduct scientific research, to access the benefits of scientific progress and to be protected against the harmful effects of science and technologies'. After tracing the historical emergence of the right (Shaver 2010), in her later work Shaver (2015) interprets the right to science as a call to frame science as a public good. This implies that 'the supply of scientific knowledge and the development of technology is must not be left entirely – or even primarily – to market forces' (Shaver 2015: 417). Shaver also proposes a 'pragmatic approach' to defining the normative content of the right, which requires being 'responsive to the particular challenges and issues of the time' (2015: 427). Using the treaty interpretation methods described in the Vienna Convention on the Law of Treaties, Donders (2011) argues that the inextricable link between the right to science and the right to health determines the positive and negative obligation of states.

Scholars have also begun investigating what implications for policymakers can be derived from applying the human rights framework to the analysis of those issues. Knoppers et al. (2014) frame their proposal for an international code of conduct enabling global genomic and clinical data sharing for biomedical research with reference to the right to science and the right to the protection of the moral and material interests resulting from scientific productions. Gran et al. (2014: 344) explore children's rights through the lens of the right to science and, after showing that indicators reveal dramatic differences in children's conditions across and within countries, argue that the right to science has the potential to address some of these inequalities by leading to 'improvements in young people's health and well-being, to greater participation in their communities, and to stronger legal protections, among other advances'. Skre and Eide (2013) connect the right to access to open access to scientific knowledge. Harris and Wyndham (2015) urged that data sharing must take into account human rights considerations, arguing that data is both a tool of scientific inquiry, to which access is vital, and a product of science, from which everyone should benefit. Vayena and Tasioulas (2015) propose using the human right to science as a promising, proper framework to develop policies in the area of citizen science.

While academic discourse has primarily developed through scholarship, various initiatives and conferences have looked at the right to science. An important initiative, focusing primarily on the United States, is led by the American Association for the Advancement of Science (AAAS). In 2009, the AAAS started a programme devoted to mobilising science and scientists to advance human rights. The project has produced important outcomes which include building a database of state reports to the UN on the implementation of the right to science (AAAS 2017a); creating the AAAS Science and Human Rights Coalition; publishing a study (authored by Margaret Weigers Vitullo and Jessica Wyndham (AAAS 2013)) on how scientists in the United States perceive the meaning and application of the right; and the Science and Human Right Report on a monthly basis (AAAS 2017b).

All of these international initiatives and scholarly works are welcome and very promising. However, in spite of all these initiatives and documents, the contours of the 'right to science' remain ill defined. It remains unclear how it should be understood, what rights society, individuals and scientists exactly have, and what the corresponding duties of the states are. To advance the debate, we argue that conceptual clarity can also be achieved via mobilisation of the right, that is to say, use of the judicial and political forums where the right can be invoked.

Mobilising the right to science: realising the right to science through judicial mobilisation

The judicial path entails bringing claims before international judicial and quasi-judicial forums against states whenever the right to science is violated. The most promising forum at the global level is the individual complaints procedure of the Committee on Economic, Social and Cultural Rights (ESCR Committee) (Forman 2016). While still in its infancy (the mechanism was established in 2008 and started operating in 2014), this tool is attractive because the committee's decisions on the right to science are, arguably, authoritative interpretations of the provision of the ICESCR (1966). These decisions will help build the body of law on this matter. The procedure requires individuals or groups of individuals to file a communication (i.e. complaint) with the committee, alleging violations of the ICESCR by a state which has ratified both the committee and the Optional Protocol (2008). Decisions made by the committee are not legally binding but, if the committee finds for the victim, its decision will contain a finding of law as well as recommendations to the state in question on how the violations should be remedied. In addition, the case will remain under consideration until satisfactory measures are taken by the state party.

Mobilisation may also involve judicial and quasi-judicial bodies at the regional level. The Court of Justice of the European Union can hear cases arising from violations of the Charter of the Fundamental Rights of the European Union. A major limitation of this process is that the Court of Justice of the European Union can only hear claims for violations committed by an institution, body, office or agency of the European Union, not by member states. Cases could also be brought in the inter-American human rights system by activating the Inter-American Commission on Human Rights, and perhaps even the Inter-American Court of Human Rights. Questions of violations of the right to science could be brought before the commission on the basis of the OAS Charter, American Declaration and Protocol of San Salvador. As in the case of the ESCR Committee, decisions of the Inter-American Commission are not binding but help build the body of law. The right to science could also be invoked before the Inter-American Court of Human Rights, a body whose decisions are binding by its interpretations of rights within the American Convention of Human Rights, over which the court has jurisdiction. (While the court's jurisdiction over the Protocol of San Salvador is limited to the right to education and the right to form trade unions, the right to science could be brought up via Article 26 of the American Convention.) Judicial mobilisation in regional bodies could target laws and regulations of states which prohibit or unreasonably restrict the freedom of scientific research, for instance as in the case of bans on research on embryos or with human–animal hybrids.[2] In addition, they could target restrictions to data sharing or access to genetic resources, and measures unreasonably limiting scientists' freedom to communicate research results, to join professional associations, to collaborate with foreign scientists or to travel internationally.[3] Some limitations are in fact acceptable as long as they 'pursue a legitimate aim, [are] compatible with the nature of this right and [are] strictly necessary for the promotion of general welfare in a democratic society' (United Nations 2012: 13, citing Article 4 of the ICESCR 1966).[4] Finally, they could challenge policies excluding marginalised populations, such as indigenous peoples, from public consultations, participation in clinical trials or membership in academia (United Nations 2012: 12).

One indirect avenue to litigate the right to science is through the right to health. The right to health is better established and recognised by more legal instruments than the right to science. A greater number of international adjudicative and quasi-adjudicative bodies can consider cases of violation of this right. Thus, besides the judicial and quasi-judicial forums listed above, the right to health can also be invoked before the European Committee of Social Rights, which reviews the Council of Europe's member states' compliance with the European Social Charter (1961, revised 1996). It should be noted that only NGOs, and only certain kinds of NGOs, and not individuals have standing before the European Committee of Social Rights. Although the European Convention on Human Rights (1950) does not guarantee a right to health, over the years the European Court of Human Rights has been called upon to consider cases having a socio-economic dimension, including health – such as questions relating to medical negligence, health and bioethics, detainees' rights, health and immigration, and health and the environment – while discussing one or more fundamental civil and political rights guaranteed under the Convention. Mobilisation advancing the right to science through the right to health could focus on policies which disregard scientific evidence in setting access to treatment and cures.

In fact, patients' right to health includes the right to access treatments and cures which are based on the best possible scientific evidence. Their human rights to health and to science would be violated, for instance, when vaccines that are proven to be both safe and effective are banned. The right to science further requires states to enable downstream use of scientific knowledge and to 'promote the transfer of technologies, practices and procedures to endure the well-being or people' (United Nations 2012: 20). The right is violated when drug or biotech companies are prohibited from developing products applying scientific knowledge.

Cases can also be brought before the African Court on Human and Peoples' Rights, which has jurisdiction over cases involving the interpretation and application of the African Charter on Human and Peoples' Rights (1981), the Protocol and any other relevant human rights instrument ratified by the states concerned.

At the global and regional levels there are several bodies which could be used by scientists, citizens and advocacy groups to challenge violations of the right to science. These 'judicial methods', however, present general challenges besides those discussed above for each procedure. First, cases can be brought only if a state party to the treaty defers to the court's or committee's jurisdiction. Not all states have ratified both the ICESCR and the Additional Protocol, creating an Individual Communications procedure. By September 2016, only twenty-six states parties had ratified both. Likewise, not all countries have ratified regional instruments and/or accepted the jurisdiction of the regional courts. Second, claimants can file a claim with a supranational body only after domestic remedies have been exhausted. This process can take several years, if not decades. Third, claims must concern the violation of the rights of one or more named individuals. Cases cannot be based on general assertions that state members are violating the right to science – for instance, by enacting a law banning certain forms of scientific research. Cases are viable only if the law interferes with the enjoyment of the right to science of a specific victim, and those victims must consent to have their case brought before an international jurisdiction. Fourth, as for any legal proceedings, complaints must be filed within a certain amount of time after the violation has occurred. Otherwise legitimate cases cannot proceed if the time frame for filing them has passed. Finally, legal proceedings in supranational courts and bodies are often slow. This is due to the number of cases brought before these bodies every year and the insufficient resources available to process them expeditiously.

Realising the right to science through political mobilisation

The second path to mobilisation is to exploit the opportunities offered by the political processes of international institutions. There are opportunities for political advocacy at both the global and regional levels. Within the United Nations, under the Universal Periodic Review (UPR), all UN member states must submit a report every five years to the UN Human Rights Council describing how they have discharged all of their international human rights obligations. The procedure covers all human rights, independently of whether the state in question has ratified any given international human rights treaty. On average, each year forty-two states are reviewed during the sessions of the UPR Working Group, which meets three times a year. Human rights experts and groups are formally recognised as important stakeholders that can submit information, which the Working Group can then use as part of its review. Experts and groups can also make statements at the regular session of the Human Rights Council, when the outcome of the state reviews is considered. Although the UPR is still looking for a precise identity (Cowan and Billaud 2015), the potential for advancing the human rights agenda is substantial. In fact, the UPR intends to provide technical assistance to states and enhance their capacity to effectively deal with human rights challenges. Even more important to the present discussion, the UPR includes a sharing of best human rights practices. With regard to the right to science, the UPR process has the potential to help further define the normative content of the right to science.

A similar mechanism is the reporting procedure of the Covenant on Economic, Social and Cultural Rights. In addition to the judicial process discussed above, the ESCR Committee reviews reports that states parties file periodically to update the committee on the what they have done to implement the Covenant. The committee examines each report, addresses its concerns and makes recommendations to the state party in the form of 'concluding observations'. As part of this process, NGOs can submit 'shadow reports' which bring to the attention of the committee facts which are relevant to their review of state parties. Representatives of accredited organisations can also attend the committee's sessions and make an oral presentation, and organise lunchtime briefings and other informal meetings during the sessions.

A broad range of less formal opportunities to contribute to shaping the human rights agenda and discussion at international bodies is also available to human rights advocates. One such opportunity is engaging the Special Rapporteur in the field of cultural rights in a discussion on the right to science. Organisations and individuals can submit reports and individual communications which point to violations of the right to science. The Rapporteur can then raise the issue with member states. In addition, human rights advocates can participate in UNESCO's discussions involving science, and in particular working groups which focus on different aspects of the right to science.

International organisations at the regional level also offer opportunities for participation in their activities. In these forums, political mobilisation takes both the form of lobbying and direct participation in working groups and debates. In Europe, science and human rights advocates can work with members of the European Parliament and those of the Parliamentary Assembly of the Council of Europe to facilitate discussions and present policy proposals promoting the right to science.

In the Americas, advocates have the option to work with various institutions. The first is the General Assembly of the Organization of American States, whose meetings are held annually. The Summit of the Americas encourages civil society representatives to participate by providing recommendations on thematic areas to the member states and assisting in the implementation of initiatives in the development of an agenda for the region. Also, the Inter-American Commission on Human Rights has created a Secretariat on Access to Rights and Equity and can establish working groups to focus on various aspects of the right to science.

In Africa, advocates can participate in the periodic reviews of the African Commission on Human Rights and the summits of the African Union. Similar to the reviews conducted by the Committee on Economic, Social and Cultural Rights, the African Union mandates that states parties submit two-yearly reports to the African Commission on Human and Peoples' Rights, describing legislative and other measures they have taken in giving effect to the rights and freedoms recognised and guaranteed by the Charter. Among them is the right to science. To this end, African Union guidelines require states parties to report on the right to science by submitting information on 'laws, administrative regulations, collective agreements and court decisions' relevant to the promotion of the right

> measures taken to ensure the application of scientific progress for the benefit of everyone . . . to promote the diffusion of information on scientific progress [and] to prevent the use of scientific and technical progress for purposes which are contrary to the enjoyment of all human rights [as well as] any restrictions which are placed upon the exercise of this right, with details of the legal provisions prescribing such restrictions. (African Union 1989)

Looking to the future of the right to science

While recognised in 1948 by the Universal Declaration of Human Rights, the right to science was almost forgotten for half a century and, to borrow from Chalmers et al. (2014), it has been 'resuscitated' only recently. As a result, the normative content of this right is not yet sufficiently clear and there is no consensus on states' duties on this right. International initiatives and scholarly work are broadening the boundaries of our understanding of what the right to science entails, and what it takes to fully realise it. All are welcome developments, but we believe that it is also paramount to work on the realisation of this right from the bottom-up, through mobilisation and use of judicial and political tools. Judicial mobilisation has the potential to promote the realisation of the right to science in two ways: by addressing and redressing specific violations of the right and by obtaining formal pronouncements of supranational bodies which contribute to defining the normative content of the right. Political mobilisation can help states focus their attention on this much-neglected right and create the space for a debate between states and between states and civil society which can promote and advance freedom of scientific research.

Mobilisation certainly faces challenges. We have discussed some procedural and political challenges. These challenges are neither negligible nor insurmountable. To promote mobilisation, it is important for academia, scientific societies and human rights advocacy groups to establish a network aimed at monitoring state actions which may be in violation of the right to science, and to develop expertise for claims based on the right to science to emerge.

Judicial and political mobilisation will not only contribute to our understanding of this right and defining its normative content, but will also ensure that states incorporate this right into their policies and respect it. The hope is that, eventually, the right to science will be fully realised. As the Special Rapporteur argued, this entails living in a world in which national laws and regulations ensure that all humans have freedom to participate to the scientific enterprise, to enjoy the benefits of science, to participate in decisions relating to science and to live in a world which fosters the development and diffusion of scientific knowledge (United Nations 2012: 19–22).

Notes

1 For further discussion of the precautionary principle, see Chapter 14 in this volume.
2 Several states prohibit the derivation of embryonic stem cell lines. According to EuroStemCell (2017), this is the case in Austria, Germany and Italy.
3 For a discussion of the right to science as the freedom to contribute to the scientific enterprise, see United Nations (2012: 12).
4 For a discussion of reasonable limitations to freedom to publish scientific data, see Chapter 6 in this volume.

References

AAAS (2017a), 'Article 15: state reports', www.aaas.org/page/article-15-state-reports (last accessed 26 August 2017).

AAAS (2017b), 'Scientific responsibility, human rights & law program', *AAAS Science and Human Rights Report*, www.aaas.org/program/scientific-responsibility-human-rights-law (last accessed 26 August 2017).

AAAS Science and Human Rights Coalition (2013), *Defining the Right to Enjoy the Benefits of Scientific Progress and Its Applications: American Scientists' Perspectives*. Report prepared by M. W. Vitullo and J. Wyndham, Washington, DC, https://mcmprodaaas.s3.amazonaws.com/s3fs-public/reports/Art15_Report_AAAS_1.pdf (last accessed 26 August 2017).

Additional Protocol to the American Convention on Human Rights in the area of Economic, Social, and Cultural Rights (1988), www.oas.org/juridico/english/treaties/a-52.html (last accessed 26 August 2017).

African Charter on Human and Peoples' Rights (1981), www.achpr.org/instruments/achpr.

African Union (1989), *Guidelines for National Periodic Reports*, www.achpr.org/files/instruments/guidelines_national_periodic_reports/achpr_guide_periodic_reporting_1989_eng.pdf (last accessed 26 August 2017).

American Declaration of the Rights and Duties of Man (1948), www.cidh.org/basicos/english/Basic2.American%20Declaration.htm (last accessed 26 August 2017).

Arab Charter on Human Rights (2004), www.humanrights.se/wp-content/uploads/2012/01/Arab-Charter-on-Human-Rights.pdf (last accessed 26 August 2017).

ASEAN Human Rights Declaration (2012), http://aichr.org/?dl_name=ASEAN-Human-Rights-Declaration.pdf (last accessed 26 August 2017).

Besson, S. (2015), 'Introduction – mapping the issues', *European Journal of Human Rights*, 4: 403–10.

Chalmers, I., Bracken, M. B., and Djulbegovic, B. et al. (2014), 'How to increase value and reduce waste when research priorities are set', *The Lancet*, 383.9912: 156–65.

Chapman, A.R. (2009), 'Towards an understanding of the right to enjoy the benefits of scientific progress and its applications', *Journal of Human Rights*, 8.1: 1–36.

Charter of the African Union (1963), www.au.int/en/sites/default/files/treaties/7759-sl-oau_charter_1963_0.pdf (last accessed 26 August 2017).

Charter of Fundamental Rights of the European Union (C 364/01) (2000), www.europarl.europa.eu/charter/pdf/text_en.pdf (last accessed 26 August 2017).

Charter of the Organization of American States (a-41) (1948), www.oas.org/en/sla/dil/docs/inter_american_treaties_A-41_charter_OAS.pdf (last accessed 26 August 2017).

Cowan, J. K., and Billaud, J. (2015), 'Between learning and schooling: the politics of human rights monitoring at the universal periodic review', *Third World Quarterly*, 36.6: 1175–90.

Donders, Y. (2011), 'The right to enjoy the benefits of scientific progress: in search of state obligations in relation to health', *Medicine, Health Care and Philosophy*, 14.4: 371–81.

European Convention on Human Rights (1950), www.echr.coe.int/Documents/Convention_ENG.pdf (last accessed 26 August 2017).

European Social Charter (1961; revised 1996), www.coe.int/en/web/conventions/full-list/-/conventions/treaty/163 (last accessed 26 August 2017).

EuroStemCell (2017), 'Regulation of stem cell research in Europe', www.eurostemcell.org/stem-cell-regulations (last accessed 26 August 2017).

Forman, L. (2016), 'Can minimum core obligations survive a reasonableness standard of review under the Optional Protocol to the International Covenant on Economic, Social and Cultural Rights?', *Ottawa Law Review*, 47.2: 561–73.

Gran, B., Waltz, M., and Renzhofer, H. (2014), 'A child's right to enjoy benefits of scientific progress and its applications', *International Journal of Children's Rights*, 21.2: 323–44.

Harris, T. L., and Wyndham, J. M. (2015), 'Data rights and responsibilities: a human rights perspective on data sharing', *Journal of Empirical Research on Human Research Ethics*, 10.3: 334–7.

Knoppers, B. M., Harris J. R., Budin-Ljøsne I., and Dove E. S. et al. (2014), 'A human rights approach to an international code of conduct for genomic and clinical data sharing', *Human Genetics*, 133.7: 895–903.

Muller, A. (2010), 'Remarks on the Venice Statement on the right to enjoy the benefits of scientific progress and its applications (Article 15 (1)(B) ICESCR)', *Human Rights Law Review*, 10.4: 765–84.

Optional Protocol to the International Covenant on Economic, Social and Cultural Rights (2008), 'List of accessions and ratifications', https://treaties.un.org/pages/ViewDetails.aspx?src=TREATY&mtdsg_no=IV-3-a&chapter=4&clang=_en (last accessed 26 August 2017).

Protocol on the Rights of Women in Africa of the African Charter on Human and Peoples' Rights (2003), www.achpr.org/files/instruments/women-protocol/achpr_instr_proto_women_eng.pdf (last accessed 26 August 2017).

Schabas, W. A. (2007), 'Study of the right to enjoy the benefits of scientific and technological progress and its application', in Y. Donders and V. Volodin (eds), *Human Rights in Education, Science and Culture – Legal Developments and Challenges*, Aldershot/Paris: Ashgate/UNESCO.

Shaver, L. (2010), 'The right to science and culture', *Wisconsin Law Review*, 1: 121–84.

Shaver, L. (2015), 'The right to science: ensuring that everyone benefits from scientific and technological progress', *European Journal of Human Rights*, 4: 411–30.

Skre, A .B., and Eide, A. (2013), 'The human right to benefit from advances in science and promotion of openly accessible publications', *Nordic Journal of Human Rights*, 31.3: 427–53.

UNESCO (2009), *Venice Statement on the Right to Enjoy the Benefits of Scientific Progress and its Applications*, http://unesdoc.unesco.org/images/0018/001855/185558e.pdf (last accessed 26 August 2017).

United Nations (2012), report of the Special Rapporteur in the field of cultural rights, Farida Shaheed, 'The right to enjoy the benefits of scientific progress and its applications', presented at the 20th session of the Human Rights Council (14 May 2012) (A/HRC/20/26), www.ohchr.org/Documents/HRBodies/HRCouncil/RegularSession/Session20/A-HRC-20-26_en.pdf.

United Nations Covenant on Economic, Social and Cultural Rights (1966), www.ohchr.org/EN/ProfessionalInterest/Pages/CESCR.aspx (last accessed 26 August 2017).

Vayena, E., and Tasioulas, J. (2015), '"We the scientists": a human right to citizen science', *Philosophy & Technology*, 28.3: 479–85.

11

The donation of embryos for research: maintaining trust

Heidi Mertes

Background

There are few areas of research that are as contentious as research on human embryos. Even within Europe, very diverse policies have been developed in regard to embryo research. Some countries – such as Germany, Ireland and Poland – strictly prohibit the destruction of embryos in research, based on the argument that embryo research violates the dignity of human life and/or conflicts with religious teachings. Other countries – such as the UK, Sweden and Belgium – not only allow, but even fund the creation of embryos explicitly for research purposes. These policies are supported by the principles of freedom of research, beneficence and proportionality, as embryo research leads to improvements in healthcare that outweigh the ethical concerns involved (Mertes 2012). Most countries have adopted a pragmatic approach, balancing the arguments mentioned above, in which the destruction of donated 'spare' embryos is allowed, but not the creation of embryos for research purposes, making the so-called discarded–created distinction (Devolder 2012). Although there is much to be said about the legitimacy of making this distinction, only research involving spare embryos – and therefore the least controversial kind of embryo research – will be discussed in this chapter. Spare embryos are embryos that have been created in the course of an IVF treatment, but will not be used for transfer in fertility treatment. This can be due to various reasons: the parental project might be abandoned or completed, certain embryos may not be eligible for transfer due to genetic defects or poor prospects of further development or transfer may no longer be possible due to legal restrictions on age at transfer or a maximum storage period of the embryos. In ideal circumstances, IVF patients are asked which disposition option they prefer for their spare embryos: donation to other infertile patients/couples, donation to research or destruction, although not all options are always offered. While there is great variance between countries, high rates of embryo donation for research purposes have been repeatedly reported (Samorinha et al. 2014).

Although many countries currently allow embryo research with these donated spare embryos, the freedom to perform research on human embryos is still under threat, as illustrated by the success of the Citizens' Initiative 'One of Us'. This initiative claimed that 'the EU should establish a ban and end the financing of activities which presuppose the destruction of human embryos, in particular in the areas of research, development aid and public health', based on the belief that '[t]he human embryo deserves respect to its dignity and integrity' (European Commission 2012). This initiative managed to gather 1,721,626 signatures from twenty-nine different European countries (mainly from Italy, Poland, Spain, Germany and Romania) and was therefore allowed a hearing at the European Parliament in April 2014. This feat has only been accomplished by two other initiatives: one campaigning against vivisection, the other for the availability of drinking water. After the hearing, the European Parliament decided not to comply with the demands of 'One of Us', but this campaign shows that embryo research remains contentious, even in countries where it is currently allowed. Therefore, it is important to maintain the trust of those who presently do support research on human embryos.

Embryo research is valuable both in clinical and in basic research. In clinical research, the most straightforward application is the optimisation of infertility treatment. When a new protocol is introduced into the clinic, for example a new cryopreservation technique, this is ideally first tested on embryos that will not be transferred and grow into a person, in order to avoid harm to future people (Dondorp and de Wert 2011). Besides this clinical research, also basic research into, for example, embryo development and human embryonic stem cell research are only possible if human embryos are made available to researchers.

Classical view on embryo research versus empirical findings

In the ethics literature on embryo research, the central issue is the moral status of the (pre-implantation) embryo. The general expectation is that those who attribute a high moral status (or even personhood) to the early embryo oppose embryo destruction and that those who attribute a low (or even no) moral status to the early embryo support embryo research (or at least do not object to it). It is therefore not surprising that many of the countries outlawing embryo research have a strong religious basis. If one believes that ensoulment takes place at conception and/or that the sanctity of human life needs to be protected, then the 'killing' of an embryo cannot be made right by referring to the benefits of the ensuing research (just as the killing of people for research purposes cannot be justified). On the other side of the spectrum, if one observes from a secular perspective that the early embryo has none of the features which might bestow on it a moral status (sentience, consciousness, rationality), then sacrificing embryos for the advancement of science and healthcare is not problematic at all.

One would expect similar considerations about the moral status of the embryo to be decisive when deliberating whether or not to donate spare embryos to embryo research. However, research into the motivations for (not) donating spare embryos to research has shown that the moral status that is attributed to spare embryos is but one factor that influences the decision whether or not to donate to research. Other, equally important factors indicative of a willingness to donate are feelings of reciprocity towards science and medicine, altruism and a willingness to help others, positive views of research in general and high levels of trust in the medical system (Samorinha et al. 2014). Also, besides the *inherent* value of the embryo that the moral status refers to, the *instrumental* and *symbolic* value that people attribute to their embryos is an important predictor of intent to donate (Provoost et al. 2009; 2012).

A first group of factors predicting the intent *not* to donate are – as expected – related to the perception of the embryo, either as a person (or more specifically a child, a brother or sister of existing children) or as a symbol of the relationship with the partner. People who attribute a high *moral* status to their embryo or a high *symbolic* status (as a symbol of the relationship between two partners) are less likely to donate embryos for research. However, even in the group of people who claim to attribute personhood to their embryos, some participants were still willing to donate them for research (Provoost et al. 2010). Besides the moral and symbolic status, also the instrumental status of the embryo was important, in the sense that many people did not want all the efforts they invested in the creation of their embryos to go to waste after their IVF treatment. A high instrumental value was therefore correlated with a higher willingness to donate embryos for research.

A second group of factors are related to a lack of trust in the researchers or a lack of information about the research projects. Specifically, people reported to be concerned about their embryos being given to other patients accidentally (Lyerly et al. 2006) or being 'grown' in the lab to a stage that they felt uncomfortable with (Provoost et al. 2010). These findings send a clear message to the research community: people are willing to donate embryos to research – sometimes even despite attributing a high moral or symbolic status to their embryos – provided they are reassured that the embryos will be used in valuable research that the donors support. In what follows, a number of elements to consider will be set out.

Information before donation: specific or general?

Openness is the first prerequisite for maintaining trust. Ideally, when someone is asked to donate embryos for research, specific information about the research project that their embryos will be used in should be provided. This is, for example, recommended by the ASRM's Ethics Committee (Ethics Committee of the American Society for Reproductive Medicine 2013) for the US and by the HFEA's Code of Practice (Human Fertilisation and Embryology Authority 2009) for the UK. However, this standard may conflict with practical considerations. In general, a preferred disposition option for spare embryos is asked of IVF patients before starting treatment. This is done to avoid having to continue storage of embryos of which the progenitors cannot be contacted, are indecisive or have deceased at a point when, for example, the maximum storage period is reached. However, although the patients can be informed about the research projects that are ongoing at the moment when they start treatment (which enables a specific consent for the donation of fresh embryos), they cannot receive any specific information about the research projects that will be conducted five years later, while their embryos may easily be stored for that length of time. In this case, there are two options: (1) only allowing research with embryos of which the progenitors can be contacted so that they can give specific consent to use their embryos in a well-defined project; or (2) ask consent for different categories of research, for example research into embryo development, stem cell research, research into genetic diseases. The second option has the advantages that it is more practical to implement for both the clinic and the researchers than recontacting the patients and that embryos that were allocated to research by the parents at the beginning of treatment are not destroyed against their wishes because they cannot be contacted to indicate the exact project for which they want to donate. A blanket consent to *any* embryo research is inappropriate and unnecessary, given the easy implementation of the second option, although potential donors are of course free to donate for all possible categories.

Besides the general research categories (e.g. embryo development, embryo implantation, cryopreservation, genetic diseases), there are at least four applications/protocols which are especially sensitive to ethical concern and for which it is therefore desirable that donors give their explicit consent: stem cell research, research into germline gene editing, research in which the embryo is extensively cultured and transfer of embryos to other researchers.

When embryos are used for the derivation of stem cell lines, the embryos themselves are destroyed, yet cells containing their DNA can be cultured for a very long period of time. One might say that although the embryo itself is destroyed, its genetic blueprint is still 'alive', which is not the case in other types of destructive embryo research. This has several implications. For example, theoretically, if a cell nucleus from that line would be inserted into an oocyte and activated, a new embryo could be created that would be almost genetically identical to the original embryo (although the mito-chondrial DNA will be that of the oocyte and epigenetic changes will be present). Although this is not a very likely application, some potential donors may find the mere possibility disturbing. Others, however, may consider it a comforting idea that the embryo they donate will (potentially) go on living in a different form. Besides the physical immortality of a stem cell line, another issue might be that researchers worldwide would gain access to these cells and that the genetic characterisation of a stem cell line could reveal information about the donors. In principle, their names will not be linked to the stem cell line, but with the advent of direct-to-consumer genetic testing services which also link different people in their databases, it is not unthinkable that a stem cell line would be traced back to a certain family. Also here, for some people this may not be an issue at all, while others may have a clear preference to donate their embryos to other types of research instead.

Also for research into genome editing, a specific consent of the embryo donors is an absolute prerequisite. The possibility of modifying the genome of human embryos has sparked calls for a moratorium (either on clinical applications or also on research applications) in the research community and sparked fears of designer babies and a return of eugenics in the general population (Baltimore et al. 2015; Lanphier et al. 2015). Given the opposition in the general population to genetically modified organisms – partly based on rational concerns over monopolies, partly based on irrational fears and the yuck factor – it is hardly surprising that genetically modified human embryos instil fear and discomfort in many. At the same time, as previously argued, genome editing is a fantastic tool in research and should therefore not be banned a priori in embryo research (Mertes and Pennings 2015; Savulescu et al. 2015). As for human embryonic stem cell research, however, only the embryos of those donors who do not have personal objections to genome editing should be used in this kind of research, regardless of whether their opposition is based on rational or irrational arguments. Disregarding donors' personal opinion on this topic, although possibly benefiting science, would be disrespectful towards donors and might undermine trust in the research community considerably, as the message will be conveyed that researchers will do 'whatever they want' with donated embryos.

A third type of embryo research for which specific consent needs to be obtained in order to maintain trust, is research in which embryos are being extensively cultured. Until recently, concerns about extensive culturing of donated embryos was unwarranted, as nobody succeeded in culturing the embryo for an amount of time anywhere near the maximum period of fourteen days. Thus, the fourteen-day limit that was recommended by the Warnock Report in 1984 and adopted by several countries, is still a common rule thirty years later. However, with recent advances suggesting that it is now feasible to culture an embryo beyond the fourteen-day limit, there have been calls to extend the limit to twenty-one days in order to be able to study stages of embryo development beyond implantation and the primitive streak (Deglincerti et al. 2016; Shahbazi et al. 2016). Without wanting to engage in this debate here, I shortly want to note that there are two possible ways of regarding the fourteen-day limit. Some regard it as an arbitrary limit that was set as a middle ground between different opinions on the ideal limit, with some people preferring a limit that allows for longer culturing, others preferring it to be more restrictive. Others regard it as a non-arbitrary limit, linked to the biological phenomenon of the primitive streak and the point at which the embryo is certain to be an 'individual' as twinning can no longer occur. The relevance of these biological facts from a moral point of view is dubious, but especially religious people tend to accord significance to them. It is therefore not improbable that some people would allow their embryos to be cultured for two, but not for three weeks. On a more general note, although there are sound scientific arguments for the extension of the fourteen-day limit, such an extension may also fuel opposition against embryo research. Reproaches of researchers going down a slippery slope and changing the rules as soon as they become obstacles are bound to be voiced and will undermine trust. A new middle ground might be found in allowing research up to twenty-one days in very exceptional cases, but this subject will undoubtedly be heavily debated.

A fourth procedure that is linked to the danger of undermining trust is transfer of embryos to other researchers and other facilities. At first glance, one would think that when an embryo is donated to research, it does not really matter whether the research is carried out in the research institution connected to the hospital where treatment was received or elsewhere. In fact, to avoid undue pressure on patients to donate and to avoid concerns regarding malpractice, the treating physician and the researcher using the embryos should not be one and the same person, which may be an argument to loosen the ties between the clinic and the research, rather than keeping them tight. However, as mentioned above, at least part of the motivation to donate embryos to research is trust in and reciprocity towards the institution where they received IVF treatment. The same relation of trust is most likely not present with other institutions performing embryo research and there may even be instances of mistrust towards other particular laboratories, for instance in the case of commercial spin-offs, or if the other institution is of a different religious background or subjected to different legislation and/or oversight. Just as for stem cell research, genome editing and extensive culturing, transfer of embryos is not necessarily problematic, but as it might be perceived as problematic in specific cases, it is better to err on the side of caution and obtain an explicit informed consent of the donors to make sure that their trust is not betrayed.

For all four of these 'ethically challenging' applications – stem cell research, genome editing, extensive culturing and transfer of embryos to other institutions – the difficulty in obtaining an informed consent for the donation will be to explain the possible issues without introducing fear (rather than alleviating fear).

Information after donation

Also, after the donation, measures can be taken to encourage trust in embryo research. Currently, there is little communication about the number of embryos donated to research each year, the number of embryos used in research each year, the goals of the research projects in which they are used, the scientific output of those projects and which types of embryos are used (fresh, frozen or – where applicable – created for research). A first report of this kind has recently been published by members of the Belgian federal commission for research on *in vitro* embryos (Pennings et al. 2017).

It is also highly desirable that the results of all studies using donated human embryos become part of the public domain. Given the sensitive nature of embryo research, given that embryo donors cannot be compensated or otherwise rewarded for their donation (although there have been calls to do this, see de Lacey 2006) and that the main motivations of embryo donors are reciprocity and the desire to help other people, their embryos should not be used to further the interests of commercial companies/spin-offs without serving the common good. If not, donors may feel that researchers are taking advantage of their altruism to further their own – non-altruistic – goals of profit-making. This does not necessarily mean that all inventions based on research in which embryos were destroyed should be unpatentable – contrary to what was decided by the European Court of Justice in the famous *Brüstle* v. *Greenpeace* case in 2011 (European Commission 2011).[1] However, the research itself, the findings about reproductive biology, embryogenesis, outcomes of different cryopreservation techniques etc. should be made public. By sharing this research, the recipient of the donated embryos in turn shows reciprocity towards the donors/IVF patients. A policy of mandatory sharing of information also prevents needless repetition of research with the valuable and scarce resource that human embryos are.

Conclusion

While the debate concerning research on human embryos is often reduced to a debate on the moral status of the embryo, several studies have found that the reason why IVF patients do or do not donate their embryos for research depends on many other factors as well. One of those factors is trust in the scientific community. Therefore, it is important that the confidence that embryo donors entrust in scientists is not betrayed. In this chapter we set out some recommendations on how to maintain this trust. A first prerequisite is that the donors are informed about the (type of) research that their embryo will (possibly) be used in. Although due to practical limitations it may not always be possible to obtain consent for the specific research project that an embryo is used in, this cannot be a reason to move to a blanket consent. A middle ground can be found by requesting consent for several categories of research, so that the patients are able to exclude types of research that they object to. For four specific applications/protocols, an explicit and specific consent is advocated: stem cell research, genome editing, extensive culturing and transfer of embryos to other research facilities. This does not pretend to be an exhaustive list, but at least for these four applications, we can imagine the possibility that even embryo donors who do not object to embryo research per se, might nevertheless object to these kinds of use of their embryos. By seeking explicit consent, we can prevent potential donors from refraining from donating to research all together from fear of one of these applications. In order to maintain the trust not only of donors, but also of the general public, transparency around the embryo research that is being performed is important. Information on how many embryos are used for which kind of projects and what the outcomes of the research are, should therefore be made public.

Note

1 Article 6(2)(c) of the EU's Directive 98/44 (Directive on the legal protection of biotechnological inventions) states that 'uses of human embryos for industrial or commercial purposes' are unpatentable as this would be 'contrary to *ordre public* or morality'. In its verdict in *Brüstle* v. *Greenpeace*, the European Court of Justice concluded that this prohibition on patenting also covers the use of human embryos in research, products whose production necessitates the prior destruction of human embryos and processes for which a base material is required which is obtained by destruction of human embryos, even if the description of the technical teaching claimed does not refer to the use of human embryos. This verdict in fact excludes the entire field of human embryonic stem cell research from patentability.

References

Baltimore, D., Berg, P., and Botchan, M. et al. (2015) 'A prudent path forward for genomic engineering and germline gene modification', *Science*, 348.6230: 36–8.

de Lacey, S. (2006), 'Embryo research: is disclosing commercial intent enough?', *Human Reproduction*, 21.7: 1662–7.

Deglincerti, A., Croft, G. F., and Pietila, L. N. et al. (2016), 'Self-organization of the in vitro attached human embryo', *Nature*, 533: 251–4.

Devolder, K. (2012) 'Against the discarded–created distinction in embryonic stem cell research', in M. Quigley, S. Chan and J. Harris (eds), *Stem Cells: New Frontiers in Science and Ethics*, London: World Scientific Publishing, 137–62.

Dondorp, W., and de Wert, G. (2011) 'Innovative reproductive technologies: risks and responsibilities', *Human Reproduction*, 26.7: 1604–8.

Ethics Committee of the American Society for Reproductive Medicine (2013), 'Donating embryos for human embryonic stem cell (hESC) research: a committee opinion', *Fertility and Sterility*, 100: 935–9.

European Commission (2011), ec.europa.eu/dgs/legal_service/arrets/10c034_en.pdf (last accessed 3 November 2017).

European Commission (2012), http://ec.europa.eu/citizens-initiative/public/initiatives/successful/details/2012/000005/en (last accessed 27 October 2017).

Human Fertilisation and Embryology Authority (2009), *Code of Practice*, London: HFEA.

Lanphier, E., Urnov, F., Haecker, S. E., Werner, M., and Smolenski, J. (2015), 'Don't edit the human germline', *Nature*, 519.7544: 410.

Lyerly, A. D., Steinhauser, K., Namey, E., Tulsky, J. A., Cook-Deegan, R., Sugarman, J., Walmer, D., Faden, R., and Wallach, E. (2006), 'Factors that affect infertility patients' decisions about disposition of frozen embryos', *Fertility and Sterility*, 85: 1623–30.

Mertes, H. (2012), 'Understanding the ethical concerns that have shaped European regulation of human embryonic stem cell research', *Proceedings of the Belgian Royal Academies of Medicine*, 1: 127–39.

Mertes, H., and Pennings, G. (2015), 'Modification of the embryo's genome: more useful in research than in the clinic', *American Journal of Bioethics*, 15.12: 52–3.

Pennings, G., Segers, S., Debrock, S., Heindryckx, B., Kontozova-Deutsch, V., Punjabi, U., van de Velde, H., van Steirteghem, A., and Mertes, H. (2017), 'Human embryo research in Belgium: an overview', *Fertility and Sterility*, 118.1: 96–107.

Provoost, V., Pennings, G., and De Sutter, P. et al. (2009), 'Infertility patients' beliefs about their embryos and their disposition preferences', *Human Reproduction*, 24.4: 896–905.

Provoost, V., Pennings, G., and De Sutter, P. et al. (2010), 'Patients' conceptualization of cryopreserved embryos used in their fertility treatment', *Human Reproduction*, 25.3: 705–13.

Provoost, V., Pennings, G., and De Sutter, P. et al. (2012), '"Something of the two of us": the emotionally loaded embryo disposition decision making of patients who view their embryo as a symbol of their relationship', *Journal of Psychosomatic Obstetrics and Gynaecology*, 33: 45–52.

Samorinha, C., Pereira, M., Machado, H., Figueiredo, B., and Silva, S. (2014), 'Factors associated with the donation and non-donation of embryos for research: a systematic review', *Human Reproduction Update*, 20.5: 641–55.

Savulescu, J., Pugh, J., Douglas, T., and Gyngell, C. (2015), 'The moral imperative to continue gene editing research on human embryos', *Protein & Cell*, 6.7: 476–9.

Shahbazi, M. N., Jedrusik, A., and Vuoristo, S. et al. (2016), 'Self-organization of the human embryo in the absence of maternal tissues', *Nature Cell Biology*, 18: 700–8.

12

From Galileo to embryos and narcotic drugs: the quest for the right to science

Marco Cappato

Introduction: from Galileo to embryos

'Foolish and absurd in philosophy, and formally heretical since it explicitly contradicts in many places the sense of Holy Scripture' (Library of Social Science 2016). This was the conclusion reached in 1615 by the Roman Inquisition on Galileo Galilei's heliocentrism. Using a telescope, Galileo had observed the moons of Jupiter and sunspots, and advocated a heliocentric solar system. He also conducted investigations in buoyancy and wrote on physics, but his research was considered a threat to established doctrines. Although the Counter-Reformation ended several centuries ago, scientific research continues to be perceived 'politically' as an activity that needs to be carried out within certain limits.

Scientific research is still often limited by seemingly 'ethical' arguments. Research on embryos, for example, has been prohibited in several countries, and restricted in others (such as the UK) in which it is legal. This restriction is not based on fears around what embryological science may find out, or fears around the applications of scientific discoveries in this area, but on allegedly 'ethical' and to some extent metaphysical or religious grounds (e.g. what it is that renders a life *a person*, and how entities that qualify as *persons* should be treated).

In this chapter I will consider the case of psychoactive substances and discuss how the ban on the personal consumption of these substances impacts on scientific research. We should note that the use of psychoactive substances is probably as old as humankind itself. Most of the oldest religions include in their rituals the use of substances that may have toxic effects on the individual: from wine to weed, from mushrooms to all sorts of alkaloids. For centuries, the relationship between the physical and the metaphysical has been accompanied by the consumption of herbs, plants and their derivatives. Those substances are also known by the generic term of 'drugs', which in English, at least, can also mean 'medicines'. In fact most, if not all, of those substances can also have a medicinal use. However widespread, many psychoactive substances are subject to an international control regime that urges governments to control their production, consumption and commerce, making these actions 'criminal offences' (art. 36 of the 1961 UN Single Convention on Narcotic Drugs – hereafter, the 1961 Convention).

Whereas it is widely accepted that psychoactive substances are used for medical purposes, or even to treat emotional disorders (e.g. psychotropic drugs are administered routinely to treat psychological and emotional disorders), allegedly moral judgement and ethical concerns, as well as legal restrictions, seem to change dramatically if, instead of seeking a cure, the 'alteration' of our consciousness is done just for the sake of it.

This chapter will focus mainly upon the prohibition of narcotics and other psychoactive substances, and its impact on science. I will discuss how international organisations, particularly the United Nations, have intervened over the years to regulate and control the use and distribution of psychoactive substances. There are three main international conventions that deal with psychoactive substances. The first is the one mentioned earlier (the 1961 UN Single Convention on Narcotic Drugs); two further conventions followed: the Convention on Psychotropic Substances 1971 and the Convention against Illicit Traffic in Narcotic Drugs and Psychotropic Substances 1988. The first, the 1961 Convention, is the one that contains the tables of plants and derivatives that must be strictly controlled. The most problematic of these conventions is the first, because it is this that lists the banned substances, and thus causes the problems that are discussed in this chapter. I will thus focus on the 1961 Convention. We will see, however, that various UN documents contain contradictions and are open to looser and stricter interpretations, thus leaving individual states uncertain as to how to regulate the use of psychoactive substances.

I will show that the prohibitions relating to psychoactive substances can seriously hinder the progress of scientific research. As scientific advancement is regarded as a human right by the same treaties and documents that restrict the use of psychoactive substances, prohibition results in the violation of fundamental human rights, such as the right to science and to health.

There is another aspect worth noting: the case of drug prohibition is illustrative of how at times rules and laws, particularly restrictive ones, are based on dogma or on unsubstantiated fears (as happened in the case of Galileo), and fail to take into consideration the scientific evidence available, rather than being grounded, as they should be, on a dispassionate analysis of the issues at stake. This method of regulation is not confined to psychoactive substances; cloning and assisted fertilisation, to highlight other examples, at least in some countries, have been similarly regulated on shaky grounds. I will begin with a brief overview of the international responses to cloning because these responses have been in many respects similar to the international reaction to psychoactive substances; this will illustrate further how scientific progress, and thus indirectly our health and welfare, are readily sacrificed in the name of empty ideals.

The case of cloning

At times, limits and prohibitions are dictated by the risks (real or perceived) of harm to someone (the research subjects, or other humans, or the environment); at other times, they appear rather to be based on more vague and even fanciful grounds. For example, as is well known, after the birth of Dolly the sheep in 1996, virtually all international organisations and institutions lined up to ban human cloning. In March 2005 the General Assembly of the United Nations finally approved a non-binding declaration calling on all UN member states to ban all forms of 'human cloning', including cloning for medical treatment, as incompatible with 'human dignity and the protection of human life' (United Nations News Centre 2005). The declaration also banned 'genetic engineering techniques that may be contrary to human dignity'. It also called on states 'to prevent the exploitation of women in the application of life sciences' and 'to protect adequately human life in the application of life sciences'. The document received the support of eighty-four out of 193 UN member states. It is important to note that this, and other similar documents, was not just a prohibition or moratorium on human cloning, but a clear expression of the ethical norms that should guide and limit scientific research.

The term 'human cloning', construed during the debate of the Sixth Committee of the General Assembly throughout 2005, included in its meaning both *therapeutic* and *reproductive* cloning. On the basis of this equivocation, some countries, such as South Africa, abstained at the United Nations in 2005, claiming that 'therapeutic cloning was aimed at protecting human life and [is not], therefore, inconsistent with the Declaration' (*The Hindu* 2005). The United States, which voted in favour of the declaration, stated that 'any ban on human cloning should explicitly state that it does not prohibit the development of cell and tissue-based therapies based on research involving cloning technology to produce DNA molecules, organs, plants, tissues, cells (other than human embryos), or animals (other than humans)' (United Nations News Centre 2005), and concluded that the United States believes that 'nations should actively pursue the potential medical and scientific benefits of these scientific methods, which have already enabled researchers to develop innovative drugs to treat diseases' (United Nations News Centre 2005).

The 2005 Declaration on Human Cloning followed the Universal Declaration on the Human Genome and Human Rights adopted by UNESCO in 1997. Both are non-binding statements. Nevertheless, the drafting process, and in particular the debate that accompanied the adoption of the declaration throughout 2004, happened at a time when the first promising results on embryonic stem cell research were being published: this demonstrates how crucial and divisive scientific research can be for decision makers.

After the adoption of the 2005 statement, the international community addressed the issue of cloning again at a technical level thanks to the work of the International Bioethics Committee (IBC) of UNESCO, which dedicated four years of its work, until 2011, to the issue of the moral acceptability of scientific research on human cloning. The IBC advised that global dialogue would greatly benefit from efforts in the three following areas:

Terminology: frameworks and regulations should not make use of inaccurate and misleading terminologies that inadequately describe the technical procedures relevant to human cloning. The new scientific developments call for the redefinition and clarification of some widely used terms and for the dismissal of others.

International governance: the IBC considered that the existing international legal frameworks and regulations were unable to properly address the challenges posed by the most recent scientific developments. They are non-binding and mutually inconsistent as a result of different views of member states. A process should be initiated that could lead to the establishment of a more robust mechanism which should give more substantial guidance on the specific issue of cloning.

Dissemination: the IBC stressed the importance of fostering public awareness by disseminating, discussing and debating cloning issues at all levels. This would allow all countries, including the developing and least developed countries, to participate in the debate and put forward their concerns regarding the new technologies related to human cloning (UNESCO 2017).

The opposition to cloning, in any form, has been fierce, in spite of the international documents welcoming *therapeutic* cloning. The problem with any type of research of this kind is that it involves the manipulation of human embryos (of course, there is a question as to whether the entity that results from so-called cloning, that is, cell nuclear transfer, is an embryo in the same way as the entity that results from the fusion of male and female gametes). But leaving the technicalities aside, the idea that humans in scientific laboratories would create, manipulate, study and possibly destroy something that looks very much like a human embryo has been found unacceptable and repugnant to many.

As is well known, *therapeutic* cloning refers to the enucleation of an oocyte and the insertion of the nucleus of an adult cell into the enucleated oocyte. With certain types of electric stimulation, an embryo can be formed (or an entity that very much looks and functions like an embryo), in absence of a fertilised egg. This embryo would have DNA that would be nearly identical to the DNA of the donor of the nucleus of the adult cell (with the exception of the mitochondrial DNA that is inherited via the oocyte). The embryo, after around five days, presents itself as a hollow cavity containing stem cells. These cells are highly *plastic*, that is, they have the ability to differentiate into many tissues, and could in principle be extracted and induced to differentiate into desired tissues, which could provide a cure for many diseases (Parkinson's disease, other neurodegenerative disorders, heart diseases and many others), free of problems of immunological rejection.

But the whole enterprise of embryological research, not only that aimed at producing 'clones', raises fierce opposition. By way of example, in 2013 a group of European organisations launched the 'One of Us' campaign to call on European institutions to unconditionally recognise 'the inherent and inalienable human dignity as a source of human freedoms and citizens' rights' (One of Us 2012–17). This 'federation' of NGOs collected 1,721,626 signatures from all over Europe to support their request to the European Commission to stop the European funding of research on embryonic stem cells. The 'citizens' initiative' (European Commission 2017) was submitted, discussed and rejected by the Commission in May 2014 (European Commission 2014). In a hearing convened to discuss the matter, Máire Geoghegan-Quinn, European Commissioner for Research, Innovation and Science, declared that

> Member States and the European Parliament agreed to continue funding research in this area for a reason. Embryonic stem cells are unique and offer the potential for life-saving treatments, with clinical trials already underway. The Commission will continue to apply the strict ethical rules and restrictions in place for EU-funded research, including that we will not fund the destruction of embryos. (European Commission 2014)

The scientific community, the regulators and civil society organisations may have been caught unprepared for 'cloning', but it is likely that the farraginous regulatory systems in place, both nationally and internationally, often guided by ideologies and dogmas rather than by dispassionate reasoning, may have set back scientific advancement and research. The therapeutic potentials of human cloning are well known, and were already disseminated soon after the announcement of the birth of Dolly; in principle, cloning could provide treatment for a number of degenerative disorders (Kfoury 2007); thus, the costs of setbacks are to be paid, in likelihood, by real people, who could, at least in principle, have benefited from scientific advancements.

Narcotic drugs: their potential in science and medicine

The more we know about how the brain works, the better we can intervene – and even alter – its functioning. Part of this endeavour is of unquestioned value, especially when we talk about serious or currently incurable diseases. Not many questioned the initiative by President Obama to invest in the so-called 'BRAIN' initiative, a 'research effort to revolutionize our under-standing of the human mind and to uncover new ways to treat, prevent, and cure brain disorders like Alzheimer's, schizophrenia, autism, epilepsy, and traumatic brain injury' (National Institutes of Health 2017). In 2013 the European Commission similarly launched the Human Brain Project (HBP) to build a collaborative scientific research infrastructure based on information and communications technologies to allow researchers across the globe to advance knowledge in the fields of neuroscience, computing and brain-related medicine (Human Brain Project 2017).

Understanding brain functioning requires its observation 'under the influence' of some substance. The substance can be either endogenous or exogenous. Research into autism in children, for instance, might look at how the brain works during exposure and interaction with significant others; this research could attempt to verify brain changes in children affected by autism when, say, their mother enters the room and interacts with them. Exposure to the mother in most cases will cause the endogenous production of chemicals in the brain of the research subjects. Other research might investigate the role of psychoactive substances in the central nervous system and thus observe brain performance, say, in heroin addicts when they take their fix. The role of researchers in both cases is, or should be, observing, analysing and researching, and not making moral judgements about people's behaviour.

The study of psychoactive drugs could be crucial to the understanding of the human brain. David Nutt, former chair of the UK Advisory Council on the Misuse of Drugs, noted that research into the human brain would benefit from the study of substances that are currently prohibited or whose use is currently severely restricted. Nutt has examined the neural effects of mind-altering drugs, such as the hallucinogen psilocybin (an active ingredient in magic mushrooms); the study led to the first images showing the effects of lysergic acid diethylamide (LSD) on the human brain, as part of a series of studies examining how the drug causes its characteristic hallucinogenic effects. In an interview with *Nature* Nutt explained that his study revealed how LSD might ultimately be therapeutically useful, reminding us how in the 1950s and 1960s, thousands of people took LSD to cure alcoholism. A retrospective analysis of some of those studies in 2012 sug-gested that the drug helped recovery from alcohol addiction. Since the 1970s there have been several research studies of LSD in animals, but not in humans, and Nutt argued that it was important to validate the trial of this drug as a potential therapy for addiction or depression in humans (Cormier 2016). Research on psychoactive substances could also prove beneficial in finding a treatment for other conditions, such as autism and post-traumatic stress disorder (Multidisciplinary Association for Psychedelic Studies 2017). If this all sounds reasonable in theory, in practice the use of psychoactive substances, as was the case with 'cloning', has been virtually universally condemned.

International and national prohibitions on the use of psychoactive substances

As mentioned in the introduction, the scheduling, that is, the restriction, of potentially medicinal plants and their derivatives is the result of the ratification of three international conventions on 'drugs' adopted by the United Nations (United Nations Office on Drugs and Crime 2017) between the early 1960s and the late 1980s. In addition to being used to justify the so-called 'war on drugs', these documents have also created the general conditions for the limitation of research into narcotics, including research for scientific or medical purposes.

The incorporation of the 1961 Convention into the various national legal systems made the production and consumption of certain psychoactive substances a criminal offence; however, the Convention did not establish a court before which countries not in compliance with its provisions could be brought for breaching them. The 1961 Convention came into force in the mid 1970s, at a time when the international community was adopting the International Covenant on Civil and Political Rights (ICCPR) (1966) and the International Covenant on Economic, Social and Cultural Rights (ICESCR) (1966), two documents that finally expanded, clarified and codified the rights enumerated in the Universal Declaration of Human Rights of 1948.

Over the years, these last two covenants were strengthened by the creation of a committee (UN Committee on Economic Social and Cultural Rights 1996–2017), which was tasked with monitoring the domestic application of the covenants. This development of a compliance mechanism is a crucial difference from the UN conventions on narcotics, and signals the hierarchical relationship between the documents. While the articles of the two human rights covenants have constitutional status, obliging states, under international law, to protect and promote the rights contained therein, those of the three narcotics conventions are a set of shared recommendations that the international community believes should be applied to ensure an 'international drug control system'; while there are two UN committees to address violations of the human rights contained in the covenants, there are no quasi-judicial bodies to assess how states enforce the norms contained in the three UN conventions on drugs. Because of their constitutional status, the rights in the covenants trump the provisions of the conventions.

The regulation of psychoactive substances has been a hard nut to crack for national governments as well. One example is provided by the saga of the regulation of cannabis in the UK. This case illustrates how even in countries where public opinion favours the legalisation of certain psychoactive substances, the response of the government may end up oscillating between decriminalisation and prohibition. In 2001, following a public survey, the British government announced that cannabis would be transferred from class B under the 1971 Misuse of Drugs Act to class C, thus decriminalising personal possession (arrest would still be possible for distribution).

Reclassification had the support of a majority of the public, with surveys at the time finding that 49 per cent of British adults supported cannabis decriminalisation, 36 per cent were against and 15 per cent were undecided. The rescheduling eventually happened in January 2004, after class C penalties for distribution had been stiffened.

The British Advisory Council on the Misuse of Drugs (1979) had already recommended such a reclassification as early as 1979, a view that was endorsed by the so-called Runciman Report in 1999 (Runciman 1999). The changes would have allowed police forces to concentrate resources on other more serious offences, including those involving 'stronger drugs'. However, in 2005 the government announced that the reclassification of cannabis from class B to class C would be reviewed in light of new scientific research, and the issue was, finally, formally referred to the Advisory Council.

In January 2006 the Home Secretary maintained that, based on advice from the Advisory Council, there would be no move to return cannabis to class B. A year and half later this decision was reconsidered. In fact, in May 2008 the government confirmed that cannabis in the UK would again be classified as a class B drug, despite the Advisory Council's recommendation. On 26 January 2009 cannabis was reclassified as a class B drug, once more making its possession and use a criminal offence.

A global war against the use of narcotics in scientific research

As the previous section has shown, national laws and international conventions virtually unanimously prohibit the recreational use of several psychoactive drugs. Whereas many of us are well aware of this prohibition, the effect that it might have on scientific research (and thus, indirectly, on human health) has not been widely recognised. In 2013 David Nutt, mentioned earlier, attempted to highlight the effects of these prohibitions on scientific research. In an article published in *Nature* he wrote:

> The possession of cannabis, 3,4-methylenedioxy-N-methylamphetamine (MDMA; also known as ecstasy) and psychedelics is stringently regulated. An important and unfortunate outcome of the controls placed on these and other psychoactive drugs is that they make research into their mechanisms of action and potential therapeutic uses – for example, in depression and post-traumatic stress disorder – difficult and in many cases almost impossible. (Nutt et al. 2013)

Similarly, Ben Sessa, in an article published in the *Lancet* in January 2015, wrote:

> For many people, words such as psychedelic and LSD (lysergic acid diethylamide) refer only to dangerous drugs of abuse . . . Less well known is that tens of thousands of patients were treated effectively with psychedelic drugs in the 1950s and 1960s and that these drugs had almost become part of mainstream

medicine by the time they became demonised and research was halted for 40 years. (2015)

Even in states in which the medicinal use of some psychoactive substances is recognised, the laws remain contradictory, and the result is an obstruction to scientific research. For example, in the United States, under federal law, twenty-eight states recognise cannabis as a medicine, and yet cannabis is treated like any other controlled substance, such as cocaine and heroin. According to its relative potential for abuse and its medicinal value, the federal government places every controlled substance in a schedule. As a matter of fact, the cannabis plant remains illegal under federal law. The federal government regulates drugs through the Controlled Substances Act (CSA) (21 USC para. 811), which does not recognise any difference between medical and recreational use of the plant. This Act is generally applied against persons who possess, cultivate or distribute large quantities of cannabis. Under the CSA, cannabis is classified as a Schedule I drug, which means that the federal government considers the plant and its derivatives as highly addictive and having no medical value. Doctors may not prescribe cannabis for medical reasons under federal law, but they can recommend it under the First Amendment. Somewhat paradoxically, the enjoyment of the right to health is possible thanks to 'freedom of speech' (Americans for Safe Access 2017).

It is easy to be appalled by some of the implications of the use and abuse of psychoactive substances (legal and illegal). Substance use claims thousands of lives every year: in 2015 in England and Wales alone, 3,674 drug-poisoning deaths were caused by both legal and illegal substances (Gayle 2016). In common discourse, illegal drugs are often associated with narco-trafficking, with disadvantaged young people caught in the net of the illegal drug trade through the promise of a better life; at the other end of the spectrum are highly privileged people, wealthy and fortunate but otherwise vulnerable, whose bodies are found lifeless after an overdose. Perhaps worst of all are the horrifying stories of toddlers being drugged and abused by their parents, or being drugged just for fun (Nauman 2016; Smith 2016).

These stories elicit sentiments of revulsion, which are so strong that it is easy to condemn psychoactive substances and support restrictive laws. The moral feelings of many suggest that abusive parents ought to be punished, that children ought to be protected and that young people in disadvantaged socio-economic contexts ought to be shielded from the exploitation of the illegal drugs trade. What is obscured, however, by the strong sentiment of revulsion is that prohibitionist policies do not effectively protect those who are vulnerable to addiction. The 'war on drugs' can end up claiming even more lives than the drugs themselves (an emblematic case is that of Mexico: a 2017 Congressional Research Service report by the United States on that country estimated that at least 80,000 people had been killed due to incidents related to organised crime since 2006. Under President Peña Nieto, overall homicide numbers have declined as much as 30 per cent, according to experts, but Mexican drug cartels take in between $19 and $29 billion annually from US drug sales (CNN 2017)). Even in Europe, where governments do not implement active 'wars on drugs' as seen in Mexico, making drugs illegal is still ineffective. The European Monitoring Centre for Drugs and Drug Addiction (EMCDDA 2016) reports that thousands of lives are claimed every year in Europe due to the use of illegal psychoactive substances, and the rates do not appear to be decreasing.

So, if prohibitionist policies and laws are meant to protect people's health and well-being, surely this means that health and well-being matter; but, if they matter, then there is a problem here, because prohibitionist policies result in an obstacle to scientific research, and this also compromises people's health and well-being. One could object that psychoactive drugs are likely to cause immediate and real harm to those who take them, whereas the benefits of scientific research on psychoactive drugs are potential and thus hypothetical, and in making a choice between reducing a real harm and producing a hypothetical benefit, it is rational to reduce the real harm. However, it is not clear that prohibitionist policies effectively protect people from the harm of psychoactive drugs, as we have seen.

Prohibitionist policies around narcotics and other psychoactive substances limit scientific research into drugs, thus hindering the understanding of how human brain reacts to certain substances, and thus how certain substances can be safely used as therapies or even recreationally.

The United Nations on narcotics and science: conflicting goals?

As we have seen, there are inconsistencies in several United Nations documents. Whereas the 1961 Convention prohibits the use of narcotics, other covenants affirm the importance of scientific progress, and state that science is *a human right*. Thus the prohibition of the use of narcotics, which impinges upon scientific advancement, jeopardises the full enjoyment of a human right.

Article 15 of the ICESCR (1966) stipulates, among other things, that the 'States Parties to the present Covenant recognise the right of everyone: (b) To enjoy the benefits of scientific progress and its applications; (c) To benefit from the protection of the moral and material interests resulting from any scientific, literary or artistic production of which he is the author'. Key to this article is the 'conservation, the development and the diffusion of science and culture'.

The United Nations Universal Declaration on Human Rights adopted on 10 December 1948 had already suggested that scientific research is a *human right*. Article 27 stated that '(1) Everyone has the right freely to participate in the cultural life of the community, to enjoy the arts and to share in scientific advancement and its benefits.' Imposing a strict system of control on scientific research into certain plants or substances without any solid justification arguably violates the human right to scientific advancement.

Consonant with the idea that scientific advancement is a human right, the ICESCR (1966: art. 15 (5)) calls on states to do their utmost to 'undertake to respect the freedom indispensable for scientific research and creative activity'. It also states that all countries party to the pact should 'recognize the benefits to be derived from the encouragement and development of international contacts and co-operation in the scientific and cultural fields' (1966: art. 15 (6)). It took twelve years to draft the version of the ICESCR that was adopted on 16 December 1966, and another decade passed before its coming into force on 3 January 1976. Over the years, 164 states have become party to the covenant, and in 2008 an Optional Protocol was concluded, which entered into force in 2013, allowing its parties to recognise the competence of the Committee on Economic Social and Cultural Rights (CESCR) to consider complaints from individuals. The committee consists of eighteen independent human rights experts that meet three times a year at the United Nations.

The ICESCR is the twin of the ICCPR, and was the result of a diplomatic compromise to satisfy the ideological approaches of the two main blocs during the Cold War – individual rights were considered crucial for the West, collective ones for the USSR and its allies. With the fall of the Soviet Union, this separation has blurred, paving the way for the drafting of other international documents that deal with health and scientific research, such as the United Nations Convention on the Rights of Persons with Disabilities, which entered into force in 2008; and, within the Council of Europe, the Convention on Human Rights and Biomedicine of 1997 (also known as the Oviedo Convention), which aims at the protection of 'Human Rights and Dignity of the Human Being with regard to the Application of Biology and Medicine', and which entered into force in 1999. Once ratified, the covenants become documents of constitutional value.

In its Preamble, the ICESCR (1966) reminds signatories that the 'individual, having duties to other individuals and to the community to which he belongs, is under a responsibility to strive for the promotion and observance of the rights recognized in the present Covenant'. Alas, as we have seen, over the last few decades, in particular in the field of health and science, 'striving for the promotion and observance' of Article 15 of the ICESCR has found potent enemies in other international treaties or declarations to control the diversion of substances to purposes other than their scientific or medical use.

The 1961 Convention contains (as arguably do all compromise documents) several contradictions and several ambiguous statements that have allowed rigid or lax interpretations in different countries. Examples of some inherent incongruences can be found in the Preamble of the document:

The Parties,
  Concerned with the *health and welfare of mankind*,
  Recognizing that the medical use of narcotic drugs continues to be indispensable for the relief of pain and suffering and that adequate provision must be made to ensure the availability of narcotic drugs for such purposes,

Recognizing that addiction to narcotic drugs constitutes a serious evil for
the individual and is fraught with social and economic danger to mankind,
    Conscious of their duty to prevent and combat this evil,
    Considering that effective measures against abuse of narcotic drugs require
co-ordinated and universal action,
    Understanding that such universal action calls for international co-operation
guided by the same principles and aimed at common objectives. [emphasis
added]

The 1961 Convention opens up with a solemn statement of concern for the
health and welfare of mankind, which among other things depends on
adequate access to essential medicines; at the same time, mankind must be
defended from the 'evil' of 'misuse' of those medicines. Whereas these
resounding statements are eye-catching, what constitutes a possible abuse
of controlled substances is arbitrarily defined. In fact, contrary to Paracelsus'
dictum that 'the dosage makes it either a poison or a remedy', there are no
annexes to the United Nations conventions on drugs to specify the quantity
of alkaloids that can be considered as intoxicating or dangerous. A decision
on the welfare of mankind was thus taken that left scientific and medical
arguments aside.

    Of course, one may question whether there is such a thing as a 'right to
science'; indeed the relationship between science, democracy and the rule
of law raises the more fundamental question of whether there is such a
thing. Boggio and Romano (Chapter 10 in this volume) have extensively
reported on this issue, and on how individual states and governments may
be able to claim further freedom in cases in which it is believed that such
a right is being violated. Mikel Mancisidor, vice-chair of the Committee
on Economic Social and Cultural Rights, argues that 'the Human Right
to Science does in fact present as a true human right in existing International
Law, even if its binding normative content still has to be clarified and better
specified' (2015). 'The Human Right to Science', he continues, 'is a right
that has been explicitly enshrined in the 1948 Universal Declaration (UD)
(art. 27) and in the 1966 International Covenant on Economic, Social and
Cultural Rights (ICESCR, art. 15). We are not therefore in any way dealing
with a new legal right' (2015).

    However, he notes that this right 'is relatively unknown even among
Human Rights experts and activists' and that it has 'generally been overlooked
by international bodies and also by states' (2015). Moreover, within the
mechanisms created by the United Nations, states often do not report
satisfactorily, as should be their obligation, on rights that, like this one,
fall under the category of 'cultural rights'. The other problem is that the
United Nations has not provided individual states with guidelines or recommendations on how to implement the ICESCR on scientific matters.

    In spite of all these limitations, the right to science is no less recognised
and important than the right to health or the right to life. One may question

the existence or moral significance of human rights altogether, but if and insofar as supranational organisations and individual states accept the value, moral and political meaning of human rights, such as the rights to health, to life, to education and so on, then they should also accept the value and moral and political meaning of other rights ratified by the same covenants, unless a significant difference were found that justified attaching lower importance to one particular human right.

It thus appears that international documents ratified to guide states in the regulation of the use of psychoactive substances incorporate conflicting or openly contradictory aims. On the one hand, they declare the importance of scientific progress for the welfare of mankind; in fact, not merely its *importance*: they explicitly state that scientific advancements and human health are fundamental *human rights*. On the other hand, they readily ban certain psychoactive substances whose use in scientific research could be vital to scientific progress on the functioning of the human brain, and thus to the protection and promotion of those same human rights.

Conclusion: research on narcotics as a concrete case for the right to science

The spectacular leap forward that the scientific world is expected to produce in the study of the human brain makes drugs, in particular those that are currently controlled by the three United Nations conventions on narcotics, one of the battlegrounds where current international and domestic policies pose a potent threat to the affirmation of the human right to science. Obviously, a right to science presupposes and implies that scientists ought to be free to do their job – in short, it presupposes a significant degree of freedom of scientific research. The case of narcotics and other psychoactive substances is one of many instances in which decision makers, through laws and policies, have imposed arbitrary limits to the freedom of scientific research, thereby hindering the full enjoyment of fundamental human rights.

Looking at what happened to Galileo, or even at recent events relating to research on human embryos, one might hope that it is only a matter of time before laws and policies will become more liberal, and research into drugs will be authorised and given the necessary public investment. Those countries that act first will have a significant advantage both in terms of public health and of economic opportunities, as it is certainly possible that the more we know about how the brain works, the more we will be able to control and ameliorate human life. The next few years will put the policymakers of the so-called Western liberal democracies to the test; their challenge is to rethink current prohibitionist policies and law, while also taking account of the fact that research into drugs does happen – most likely in the hands of organised crime, which constantly produces ever more powerful and more addictive substances, purposely created and distributed simply to maximise its own profit.

The danger of not loosening regulations on scientific research into controlled substances is therefore wide-ranging. It increases the chances of illegal trade, of illegal and dangerous research at the hands of dubious companies or organised crime, and it is an impediment to crucial discoveries that could lead to effective treatments for many diseases. In addition to this, there is also a general risk that democracy will suffer a heavy defeat, as overly restrictive national policies on drugs put into question the real nature of the democratic liberal model based on civil liberties (on the relationship between democracy and scientific freedom, see Chapter 13 in this volume).

Strengthening international instruments and mechanisms to affirm the right to science, and to free up the use of narcotics and other psychoactive substances in respectable scientific laboratories, could thus contribute both to science and to the protection of human rights, democracy and the rule of law.

References

Advisory Council on the Misuse of Drugs (1979), *Report on a Review of the Classification of Controlled Drugs and of Penalties under Schedules 2 and 4 of the Misuse of Drugs Act 1971*, London: Home Office.

Americans for Safe Access (2017), 'Federal marijuana law', www.safeaccessnow.org/federal_marijuana_law (last accessed 26 October 2017).

CNN (2017), 'Mexico's drug war: fast facts', http://edition.cnn.com/2013/09/02/world/americas/mexico-drug-war-fast-facts (last accessed 26 October 2017).

Congressional Research Service (2017), *Mexico: Organized Crime and Drug Trafficking Organizations Report by Beittel, J. S.*, https://fas.org/sgp/crs/row/R41576.pdf (last accessed 26 October 2017).

Controlled Substances Act (1970), www.fda.gov/regulatoryinformation/lawsenforcedbyfda/ucm148726.htm (last accessed 26 October 2017).

Convention for the Protection of Human Rights and Dignity of the Human Being with regard to the Application of Biology and Medicine: Convention on Human Rights and Biomedicine (Oviedo Convention) (1997), www.coe.int/en/web/bioethics/oviedo-convention (last accessed 26 October 2017).

Cormier, Z. (2016), 'Brain scans reveal how LSD affects consciousness', *Nature*, www.nature.com/news/brain-scans-reveal-how-lsd-affects-consciousness-1.19727 (last accessed 26 October 2017).

EMCDDA (2016), *European Drug Report: Trends and Developments*, www.emcdda.europa.eu/system/files/publications/2637/TDAT16001ENN.pdf (last accessed 26 October 2017).

European Commission (2014), 'Press release: European Citizens' Initiative: European Commission replies to One of Us', http://europa.eu/rapid/press-release_IP-14-608_en.htm (last accessed 26 October 2017).

European Commission (2017), 'The European Citizens' Initiative: official register', http://ec.europa.eu/citizens-initiative/public/welcome (last accessed 26 October 2017).

Gayle, D. (2016), 'Drug-related deaths hit record levels in England and Wales', *The Guardian*, 9 September, www.theguardian.com/society/2016/sep/09/drug-related-deaths-hit-record-levels-england-wales (last accessed 26 October 2017).

The Hindu (2005), 'Cloning incompatible with human dignity: U.N.', www.thehindu.com/2005/03/10/stories/2005031004951400.htm (last accessed 26 October 2017).

Human Brain Project (2017), 'Science overview', www.humanbrainproject.eu/en/science/overview (last accessed 26 October 2017).

International Covenant on Civil and Political Rights (1966), www.ohchr.org/en/professionalinterest/pages/ccpr.aspx (last accessed 26 October 2017).

International Covenant on Economic, Social and Cultural Rights (1966), www.ohchr.org/EN/ProfessionalInterest/Pages/CESCR.aspx (last accessed 26 October 2017).

Kfoury, C. (2007) 'Therapeutic cloning: promises and issues', *McGill Journal of Medicine*, 10.2: 112–20.

Library of Social Science (2016), 'Galileo and truth', www.libraryofsocialscience.com/newsletter/posts/2016/2016-05-12-Galileo.html (last accessed 26 October 2017).

Mancisidor, M. (2015), 'Is there such a thing as a human right to science in international law?', *European Society of International Law*, 4.1: 1–6, www.esil-sedi.eu/sites/default/files/Mancisidor%20Reflection%20%28Word%29.pdf (last accessed 26 October 2017).

Misuse of Drugs Act (1971), www.legislation.gov.uk/ukpga/1971/38/contents (last accessed 26 October 2017).

Multidisciplinary Association for Psychedelic Studies. (2017), 'MDMA-assisted psychotherapy', www.maps.org/research/mdma (last accessed 26 October 2017).

National Institutes of Health (2017), 'What is the brain initiative?', https://braininitiative.nih.gov/index.htm (last accessed 26 October 2017).

Nauman, Z. (2016), 'Mummy gave me sleep juice. Parents charged with injecting their three kids with heroin telling them "it's feel-good medicine" at home infested with rat droppings', *Sun*, www.thesun.co.uk/news/2096324/parents-charged-injecting-kids-heroin-feel-good-medicine-home-rat-droppings (last accessed 26 October 2017).

Nutt, D. J., King, L. A., and Nichols, D. E. (2013), 'Effects of Schedule I drug laws on neuroscience research and treatment innovation', *Nature Reviews Neuroscience*, 14.8: 577–85, www.nature.com/nrn/journal/v14/n8/abs/nrn3530.html (last accessed 26 October 2017).

One of Us (2012–17), www.oneofus.eu (last accessed 26 October 2017).

Runciman, R. (1999), *Report of the Independent Inquiry into the Misuse of Drugs Act 1971*, Police Foundation, www.druglibrary.org/Schaffer/library/studies/runciman/default.htm (last accessed 26 October 2017).

Sessa, B. (2015), 'Turn on and tune in to evidence-based psychedelic research', *The Lancet Psychiatry*, 2.1: 10–12, www.thelancet.com/pdfs/journals/lanpsy/PIIS2215-0366%2814%2900120-5.pdf (last accessed 26 October 2017).

Smith, O. (2016), 'Watch: twisted adults film themselves blowing marijuana smoke in toddler's face', *Express*, www.express.co.uk/news/world/703547/video-adults-film-marijuana-smoke-toddler-s-face-shocking (last accessed 26 October 2017).

UNESCO (2017), 'Human cloning and international governance', www.unesco.org/new/en/social-and-human-sciences/themes/bioethics/international-bioethics-committee/ibc-sessions/eighteenth-session-baku-2011/human-cloning (last accessed 26 October 2017).

United Nations Committee on Economic, Social and Cultural Rights (1996–2017), 'Committee on Economic, Social and Cultural Rights', www.ohchr.org/en/hrbodies/cescr/pages/cescrindex.aspx (last accessed 26 October 2017).

United Nations Convention against Illicit Traffic in Narcotic Drugs and Psychotropic Substances (1988), www.unodc.org/pdf/convention_1988_en.pdf (last accessed 26 October 2017).

United Nations Convention on Narcotic Drugs (1961), www.unodc.org/pdf/convention_1961_en.pdf (last accessed 26 October 2017).

United Nations Convention on Psychotropic Substances (1971), www.unodc.org/pdf/convention_1971_en.pdf (last accessed 26 October 2017).

United Nations News Centre (2005), 'General Assembly approves declaration banning all forms of cloning', www.un.org/apps/news/story.asp?NewsID=13576

Universal Declaration on the Human Genome and Human Rights (1997), www.unesco.org/new/en/social-and-human-sciences/themes/bioethics/human-genome-and-human-rights

Universal Declaration of Human Rights (1948), www.un.org/en/universal-declaration-human-rights (last accessed 26 October 2017).

United Nations Office on Drugs and Crime (2017) 'Treaties', www.unodc.org/unodc/en/treaties/#Drugrelated (last accessed 26 October 2017).

13

Science, self-control and human freedom: a naturalistic approach

Gilberto Corbellini and Elisabetta Sirgiovanni

A recurring assumption among political philosophers is that freedom as the ancients conceived it was different from the kind of freedom experienced in the modern world. On 13 February 1819, in his famous lecture on *The Liberty of the Ancients Compared with that of the Moderns* held at the Athénée Royal in Paris, Benjamin-Henri Constant de Rebecque gave one of the most brilliant formulations of liberal thought. Constant affirmed that modern men's liberty is 'individual liberty', whereas that of the ancients was the freedom to collectively exercise sovereignty. He wrote that the ancients were 'machines, whose gears and cog-wheels were regulated by the law'. The moderns, after Hobbes, Locke, Spinoza and Hume, used the law to circumscribe the space of expression of autonomy, which was intended a priori as indefinite and – this is the crucial difference from the ancients – no longer located in the public forum but in the so-called 'inner' or individual conscience. So, the function of law had changed, as it no longer prescribed what a citizen *must* do but what he or she *may* do. In one phrase, popular in English law: everything that is not forbidden is allowed. In this sense, the law becomes the premise, in Constant's words, for the 'pacific enjoyment of private independence'.

Constant was presenting the classical liberal theory of freedom, in which freedom is understood as the right to be subject solely to the law, and therefore not to be arrested, imprisoned, sentenced to death or mistreated by the arbitrary will of one or more individuals. Moreover, liberal freedom implies the right to express one's opinion, to choose one's work and to perform it, to make use and abuse of one's own private property, to associate with those one prefers and to exercise an influence on the administration of government. Of course, traces of such modern ideas of liberty can be found in antiquity, but wherever slavery was legal, freedom was thought of in opposition to slavery, and since the subordination of the individual to the state and the laws had to be total, the idea of 'negative liberty' found little room to advance. Even for Aristotle, the philosopher who comes closest to a modern way of thinking, what distinguished the slave from the free man in the *Politics* was not that the former was limited in his actions or subject to coercion, while the latter enjoyed personal autonomy, but the fact that everything a slave did was done to serve someone else's interest.

One question to which neither Constant nor others have provided satisfactory answers remains, however. What really happened to catalyse the cultural processes from which the moderns' ideas of liberty were shaped? In this chapter, we would like to provide a naturalistic answer to this question.

Being a naturalist about human behaviour means believing that human behaviour, including moral and social behaviour, results from the functioning of the human brain, a complex, multilevel, anatomical organ evolutionarily and developmentally determined by genetic and epigenetic factors, and that understanding our social brains is a crucial step towards explaining social facts (Corbellini and Sirgiovanni 2013). Throughout the history of philosophy, naturalists have been most concerned about the existence and sources of freedom, since while they need to face the problem of reconciling freedom with determinism just as spiritualists do (for whom it is not nature that determines human behaviour, but God or some other spiritual force), naturalists cannot escape the problem by appealing to an immaterial soul or mind. We will not deal with these metaphysical questions here, but we will examine the evolutionary and historical processes that allowed human beings to achieve those skills that determined the relevant cognitive and moral improvements that we today define as 'freedom', those that made liberal and democratic achievements possible. We believe that the introduction of modern science, especially experimental method, had a crucial role in the process (Corbellini 2011).

Naturalistic hypotheses about science and human freedom

Curiously enough, historical accounts of human liberty give very marginal consideration to the role played by science in the birth of modernity, particularly in the construction of the epistemological and psychological conditions for the development of the kind of behaviour we today consider compatible with the civic and democratic coexistence of individuals. If it is unquestioned that civic and democratic behaviour was inspired by the introduction of a new idea of freedom, that is, autonomy or self-determination, including both positive and negative components of liberty (respectively, liberty as the capacity to self-govern and liberty as the absence of external interferences or constraints), there is far less agreement about the role that modern science had in this cultural process. Important exceptions to this tradition are books by Niall Ferguson, especially *Civilization: The West and the Rest* (2011), and the volume *The Science of Liberty: Democracy, Reason and the Laws of Nature* (2010) by the science writer Timothy

Ferris. As they interestingly argue, the rise of the modern idea of liberty, along with the evolution of the psychological capabilities that permit much more advanced possibilities for planning and control, has been a consequence of the advent of experimental science and thus of the spread of critical thinking applied in a systematic way. The pioneers of political liberty were well equipped with the scientific knowledge of their time – a quite obvious claim. The idea that modern science was at the origin of the democratic revolution rests on the fact that a sizeable proportion of scientific minds were directly involved in the revolutions that led to the establishment of fundamental human rights and therefore to the birth of modern democracies, immediately before the Enlightenment and during the scientific revolution. More compelling is the reason why science made this contribution, which is, according to Ferris, that science *required* liberty and so produced social benefits by creating a sort of symbiotic relationship between liberty and science, in which the freer nations were better able to move forward with scientific endeavour, which in turn rewarded populations with knowledge, well-being and power.

Among the particular characteristics of scientific endeavour, Ferris notes that science is anti-authoritarian, self-correcting, requires the production of specific intellectual resources, is powerful in the action of transforming nature and is a social activity. These are characteristics long observed by scientists, philosophers and sociologists during the period in which science was a model of knowledge. In addition, according to thinkers who are politically very different from one another – such as John Dewey, Michael Polanyi, Joseph Needham and Karl Popper – science and democracy share epistemological and ethical–political aspects. In particular, science and democracy both require tolerance, scepticism, rejection of authority, respect for facts, freedom of communication and free access to results.

More specifically, the discovery of the advantages of the hypothetical–deductive and experimental method, starting from the ancient traditions of logical and naturalistic thought, entailed the exercise and improvement of abstract and metacognitive reasoning capacities. This determined a more decisive detachment of naturalistic research from common sense at the dawn of the modern age. Theories that scientifically explain natural phenomena are almost always counter-intuitive. We might mention a very long list: heliocentrism, the Galilean–Newtonian theory of motion, the statistical-mechanical theory of heat, the Darwinian theory of evolution, the Mendelian theory of inheritance, the second law of thermodynamics, the restricted and general theory of relativity, quantum mechanics, the mathematical theory of communication (information theory), neurobiological theories of memory and leaning, cognitive and evolutionary theories of psychology, along with many others. Abstract thought had been used by Greek philosophers and naturalists and had spread for centuries, also being cross-fertilised by other oriental traditions, but only when it met with a set of material conditions such as the ecological diversity of the economic productive systems of the late Middle Ages could it stimulate the contagious development of counter-intuitive explanations through brain training for 'unnatural' thought.

Starting to think like a scientist was a new and powerful resource for real-life problems in the modern world. By using analysis, abstraction, hypothetical reasoning and experimental control, a series of evolutionary predispositions, which were not set spontaneously, were triggered. What were the neural structures involved?

From science to freedom through self-control

By looking at neurocognitive research about scientific learning, we may infer which brain structures were crucially trained and enhanced by science so as to determine the birth of the concept of liberty in the modern age. As we know, science learning activates frontal and prefrontal brain areas to a great extent, areas that are responsible for cognitive processes, affective regulation and social choices and contribute both to the capacity for control over the environment (e.g. the ability to find or produce food, to reproduce more and with a better chance of survival for the offspring, to treat illnesses and reduce suffering) and self-control (the effortful capacity to regulate one's automatic thoughts, feelings and behaviour in favour of long-term interests or goals).

Clinical and experimental neuroscience show that frontal areas, which are also the last to reach complete maturity at the end of adolescence (in the period therefore during which the acquisition of scientific thinking can take place), are implicated in abstract reasoning, calculation and information processing for planning one's own behaviour. These capacities adaptively improve individuals' goal-directed actions and behavioural performance. From primatology and anthropological research, we may suppose that over hundreds of thousands of years it was human beings' constant need to meet practical challenges that represented the main selective pressure for the progressive evolution of these complex structures. Presumably beginning from the fabrication of lithic artefacts, which in the course of millennia went through increasingly developed symmetries and forms, higher and higher capacities of mental representation for conceiving implementable models of artefacts were selected. Other authors have extended this hypothesis to social needs. If we consider that the increase in primates' brain size and capacity is not explicable solely by daily practical activities such as finding food or defence from predators, but is much more conceivable as an adaptive response to the increasingly complex social environments in which they evolved (Byrne and Whiten 1988; Dunbar 1998; Whiten 2000; Whiten and van Schaik 2007), it seems that the development of intelligence was a consequence of the extension of social groups and of the growth and development of new kinds of relationships among their members, apart from parental or sexual bonding. New interpersonal skills were required, such as the necessity to create alliances and to negotiate, to scrutinise others or manoeuvre them, in order to give rise to increasingly complex and relatively stable hierarchical social structures. From the Neolithic period, our ancestors must have possessed roughly the same neural structures as us, which they used to free themselves from material and social conditioning and to manage social group hierarchies.

We can suppose that historical changes intervened in the expression and potentialities of such structures, whose most powerful enhancement came with the introduction of scientific method in the modern age, which we believe involves two types of intelligence, contrary to what is generally claimed. The first is a strictly academic intelligence, estimated by conventional psychometric measures of so-called crystallised (gC) and fluid (gF) general cognitive intelligence, according to the standard Cattel–Horn–Carroll (CHC) theory (McGrew 2005). gC describes the cognitive functioning detected by classic psychometric methods, such as for example IQ tests, and based on previously acquired knowledge available in long-term storage (e.g. semantic knowledge, episodic memory), while gF refers to more flexible reasoning capacities such as problem-solving ability, the ability to see relationships and analogies, to abstractly represent concepts and figures, or to combine letter or number series – all skills which are independent from prior experience and learned knowledge.

The second kind of scientific intelligence is a social intelligence, that is, the capability to navigate and negotiate with others, because scientific method is an intersubjective process that depends upon the careful inspection of results, as well as their approval and sharing by other members of the scientific community. Socially speaking, science is a liberal–democratic system in which scientists have freedom of thought and expression and share values and rules that allow rational comparison between divergent options in order to harmonise choices by different individuals, whose expectations rarely coincide. In this sense, science requires the employment of social capacities that depend on cognitive capacities such as mindreading (i.e. the capacity to attribute, explain and predict others' mental states) and emotional resonance, not only strictly academic capacities, so as to protect and restrict the concept of freedom in order to impede possible infringements but also to avoid negative outcomes as chaos.

Needless to say, scientific reasoning and practice modify the brain plastically and evidently increase cognitive skills. Neuroimaging findings show that, even if different neural networks distributed across the entire brain probably contribute to intelligence, aspects of both the so-called crystallised intelligence and fluid intelligence, especially working memory, attention, complex decision-making processes and high-level behaviour planning, are integrated in frontal areas, which are characterised by increased connectivity in the brain matter, thanks to which information processing is incredibly enriched. Nowadays there is great interest in fluid cognitive capacities, as neuroscience has still to explain why these capacities show much more age-related decline in comparison with crystallised ability (Deary et al. 2010). Although it is true that research in psychology has found evidence that social intelligence is a relatively distinct domain from general cognitive intelligence, because the former involves much more emotional processes (Kihlstrom and Cantor 2000), there are also important overlapping systems, such as those devoted to problem-solving abilities, knowledge, memory, language or creativity, sustained by the activity of the prefrontal cortex, which is also presumed to function by connecting and integrating the activities of frontal systems for reasoning and limbic ones for emotions in a much more interactive brain architecture (Barbey et al. 2014). Nonetheless, as research into different disorders such as autism or psychopathy has shown (Greenspan and Love 1997; de Oliveira-Souza et al. 2016), social intelligence is not merely a matter of planning actions but also comprises capacities such as theory of mind and emotional resonance and affective response to others, and hence it also involves the activity of the temporal lobes and limbic areas. A feature of social competence is the masterly regulation between rational decision-making and emotional impulses. The notion of 'self-control' encompasses a broad range of diverse and complex behaviour of the sort that this regulation produces.

In neurobiology, the capacity for self-control or self-regulation has been correlated to the functioning of the so-called reward system, a widely distributed network of brain structures (fronto-basal-ganglia circuit) (Aron et al. 2007), which includes the dynamic interaction between frontal and limbic areas that respectively underlie rational and emotional processes, and is regulated by the projection of neurotransmitters such as dopamine, norepinephrine and serotonin. Extended research (see Hassin et al. 2010) has shown that the development of these capacities depends on the combination of genetic dispositions and environmental triggers such as upbringing, education and relationships with peers. A long series of neurological and psychiatric pathologies (such as OCD, addiction, eating disorders or psychopathy, to mention some) have been studied as self-control deficits and associated with failures in cognitive and social domains, while psychological studies have highlighted that developing self-control competence is widely beneficial, since it contributes to personally and socially positive outcomes in life such as well-being, autonomy, and career and interpersonal success (Tangney et al. 2004).

Something that we want to point out here is that we have reason to believe that the introduction of modern science was crucial to the enhancement of the functioning of the prefrontal cortex, resulting not only in an increase in academic competence but also contributing to the enhancement of other abilities in the social domain, mainly the wise understanding and managing of the interpersonal situations and transactions required to achieve collectively desired goals, which is usually referred to as social intelligence. These empowering effects on the capacity for self-control generated the bases for experiencing and conceptualising the modern idea of liberty.

Some might find the link we are making between the exercise of self-control and freedom bizarre, and the most common reason is that in ordinary thinking, self-control is wrongly identified with obedience to authority and social conformity, with examples coming from military or educational contexts where self-controlled soldiers and students are those who show total observance of and compliance with orders coming from higher authorities. However, the proper expression of the capacity to alter or override dominant response tendencies – self-control, as described in neuropsychological literature (Mischel 2014) – for example, deferring gratification and exercising impulse inhibition, such as resisting a delicious fattening cake or remaining calm in the face of one's aggressive boss – includes the agents' fundamental competence to understand and counteract the harmful effects of situations so as to regulate goal-directed actions, even interpersonally, and not to produce automatic responses, which submission and conformity reactions typically are. This means that, even if influenced by habit and automaticity, self-control capacity allows agents to adjust their behaviour according to their own values and chosen commitments, showing the best expression of wise self-government in social domains.

In this sense, we intend self-regulation to go along with willpower (Baumeister and Tierney 2011). Even if contemporary neuroscience has clarified that much of the will is expressed outside the agents' awareness (Roskies 2012), this is not really something that concerns us here, since unconscious willpower is sufficient to self-determination, autonomy and critical thinking, the kind of capacities we believe that science education is able to enhance.

Science education and self-control enhancement for social improvement

The best lesson we can learn from neurocognitive theories of control concerns the effects of basic cognitive mechanisms of self-regulation on intrapersonal and societal-level goals and motivations. As we argued in the previous section, cognitive and social intelligence are relatively distinct but interacting capacities, and both are required for better academic achievements. What it is crucial to understand, however, is that neural mechanisms connecting rational decision-making and affect so as to exercise self-control and self-regulation are those which are enhanced by science education. First, the learning of a way of thinking that is unnatural, in the sense that it is not developed spontaneously or necessarily, makes it possible to practise 'unnatural', complex, social behaviours, such as those that lead to economic exchanges (free market) or to democracy, where decisions are made by cultivating rational discussion rather than relying on emotional impulses, and where the personification of authority as an absolute guarantee of rules of social control is no longer needed. Each individual becomes capable of self-regulation, making reference to a law applied in an impartial way (the rule of law) within the context of a political system that can now become representative. Second, in order to coach students to succeed in research, complex scientific training shapes their neurocognitive processes in a way that makes them aware of the benefits of self-control and self-regulation, such as more solid and long-term rewards and gratification in their work and lives.

There is an empirical indicator of this social change, the so-called Flynn effect (from the name of the academic James Flynn), which has detected a surprising increase in IQ of about three points every decade in Western developed countries, starting from the twentieth century (Flynn 2009). Authors such as Stephen Pinker (2011) have brilliantly correlated this to sociological research showing a decrease in violent criminal behaviour, and to the development of liberal democracies in those countries. According to Michael Shermer (2015), a great part of this phenomenon is attributable to the spread of science education in the general population, especially to abstract, categorical and hypothetical reasoning along with self- and other-mindreading, which determined the development of kinds of intelligence and behaviour modelled on the functioning of scientific processes.

Science goes beyond the mere formulation of hypotheses by adopting a criterion for the empirical testing of the control statements, which may be false. In this way, even though this may be unpleasant, scientists realise that they can always be wrong. Scientific training over time teaches us how to avoid the types of mental shortcuts or biases that systematically lead to wrong inferences and conclusions, stimulating critical and sceptical thinking unrestricted by external conditioning and governed by one's own scrutiny, or, better, an implicit model of both honest and autonomous behaviour which resembles the kind conceptualised by modern liberty. The intrinsic social nature of science, whose actors must respond to the rigorous examination of a skilled community but possess the tools and skills to prove an alternative and revolutionary theory to be true – a form of social expertise that strategically enables scientists to achieve goals associated with the mission of science – is the kind of ground that allows responsible behaviour to flourish in societies.

Some might object that episodes of misconduct and malpractice among scientists seem to show that the scientific community is not always the best example of ethical behaviour, and it is obviously true that much work in this sense must be done. Nevertheless, it should be clear that committing acts such as data falsification, fabrication or plagiarism, as well as workplace bullying, is a clear violation of science rules, and scientists who are shown to do this risk being marginalised and punished. Hence this is not the kind of behaviour that science promotes. Rather, identifying scientific work with these bad episodes has caused very negative outcomes in the past, from the stigmatisation of scientists and mistrust on the part of the general population to wrong political choices that have favoured the growth of ascientific and magical thought. There are many possible solutions to scientific misconduct, which mostly concern the need for governmental reforms to improve the working conditions of scholars in contemporary academia around the world, since unethical behaviour is prevalently the result of career pressure and reputation seeking, since academics both old and young face an insufficiency of financial resources to conduct their research, an objective impediment to the expression of their scientific freedom that has consigned academia to one of its worst periods. Unfortunately, a more detailed examination of these circumstances goes beyond the purpose of this chapter. Nevertheless, these considerations align again with empirical research into self-control and autonomy, which shows that proper expressions of self-regulation and freedom, although emerging properly in competitive contexts, are possible only in non-stressful and resource-consuming environments, where hard effort is rewarded accordingly (Baumeister et al. 1998; Gino et al. 2011; Pyone and Isen 2011).

Occasionally a certain naivety might also be present among scientists, resulting from the fact that many of them fulfil their duties and interpret their professional roles as mere 'puzzle-solvers' in the famous Thomas Kuhn (1970) sense, meaning that they conduct their research mechanically within accepted standards, as if solving crosswords or playing chess or completing jigsaws, and are very rarely dedicated to revolutionary or extraordinary investigations, being also often unaware of the epistemological dynamics of the scientific research they are involved in. This is properly the educational task that pertains to empirically informed human and social sciences, which are specifically called to sustain and complete scientific education so to train future generations to be aware of and disclose epistemological biases, and be engaged in the effort of correcting them profitably for their own best interests and for the interests of others.

References

Aron, A. R. et al. (2007), 'Converging evidence for a fronto-basal-ganglia network for inhibitory control of action and cognition', *Journal of Neuroscience*, 27: 11860–4.

Barbey, A. K., Colom, R., and Grafman, J. (2014), 'Distributed neural system for emotional intelligence revealed by lesion mapping', *Social Cognitive and Affective Neuroscience*, 9.3: 265–72.

Baumeister, R. F., Bratslavsky, E., Muraven, M., and Tice, D. M. (1998), 'Ego depletion: is the active self a limited resource?', *Journal of Personality and Social Psychology*, 74.5: 1252–65.

Baumeister, R. F., and Tierney, J. (2011), *Willpower: Why Self-control is the Secret to Success*, New York: Penguin.

Byrne, R. W., and Whiten, W. (1988), *Machiavellian Intelligence: Social Expertise and the Evolution of Intellect in Monkeys, Apes and Humans*, Oxford: Oxford University Press.

Corbellini, G. (2011), *Scienza, quindi Democrazia*, Turin: Einaudi.

Corbellini, G., and Sirgiovanni, E. (2013), *Tutta colpa del cervello. Introduzione alla neuroetica*, Milan: Mondadori Education.

Deary, I. J., Penke, L., and Johnson, W. (2010), 'The neuroscience of human intelligence differences', *Nature Reviews Neuroscience*, 11.3: 201–11.

De Oliveira-Souza, R., Zahn, R., and Moll, J. (2016), 'The neuropsychiatry of moral cognition and social conduct', in S. M. Liao (ed.), *Moral Brains: The Neuroscience of Morality*, New York: Oxford University Press, 203–36.

Dunbar, R. I. M. (1998), 'The social brain hypothesis', *Evolutionary Anthropology*, 6: 178–90.

Ferguson, N. (2011), *Civilization: The West and the Rest*, New York: Penguin.

Ferris, T. (2010), *The Science of Liberty: Democracy, Reason and the Laws of Nature*, New York: HarperCollins.

Flynn, J. R. (2009), *What Is Intelligence. Beyond the Flynn Effect*, New York: Cambridge University Press.

Gino, F., Schweitzer, M. E., Mead, N. L., and Ariely, D. (2011), 'Unable to resist temptation: how self-control depletion promotes unethical behaviour', *Organizational Behaviour and Human Decision Processes*, 115: 191–203.

Greenspan, S., and Love, P. F. (1997), 'Social intelligence and developmental disorder: mental retardation, learning disabilities, and autism', in E. MacLean (ed.), *Ellis' Handbook of Mental Deficiency: Psychological Theory and Research*, Mahwah, NJ: Erlbaum, 311–42.

Hassin, R. R., Ochsner, K. N., and Trope, Y. (eds) (2010), *Self-Control in Society, Mind, and Brain. Oxford Series in Social Cognition and Social Neuroscience*, New York: Oxford University Press.

McGrew, K. S. (2005), 'The Cattel–Horn–Carroll theory: past, present and future', in D. P. Flanagan and P. L. Harrison (eds), *Contemporary Intellectual Assessment: Theories, Tests, and Issues*, 2nd edn, New York: Guilford Press, 136–82.

Mischel, W. (2014), *The Marshmallow Test: Mastering Self-Control*, New York: Hachette.

Kihlstrom, J. F., and Cantor, N. (2000), 'Social intelligence', in R. J. Sternberg (ed.), *Handbook of Intelligence*, 2nd edn, Cambridge: Cambridge University Press, 359–79.

Kuhn, T. (1970 [1962]), *The Structure of Scientific Revolutions*, Chicago: University of Chicago Press.

Pinker, S. (2011), *The Better Angels of Our Nature: Why Violence Has Declined*, New York: Viking Books.

Pyone, J. S., and Isen, A. (2011), 'Positive affect, intertemporal choice, and levels of thinking: increasing consumers' willingness to wait', *Journal of Marketing Research*, 48: 532–43.

Shermer, M. (2015), *The Moral Arc: How Science Leads Humanity toward Truth, Justice and Freedom*, New York: Henry Holt.

Tangney, J. P., Baumeister, R. F., and Boone, A. L. (2004), 'High self-control predicts good adjustment, less pathology, better grades, and interpersonal success', *Journal of Personality*, 72: 271–322.

Whiten, A. (2000), 'Social complexity and social intelligence', *Novartis Foundation Symposium*, 233: 185–96.

Whiten, A., and van Schaik, C. P. (2007), 'The evolution of animal "cultures" and social intelligence', *Philosophical Transactions of the Royal Society of London. Series B, Biological Sciences*, 362.1480: 603–20.

14

Evidence-based policy and the precautionary principle: friends or foes?[1]

Roberto Baldoli and Claudio M. Radaelli

A key theme in our volume is the connection, or, most pertinently perhaps, disconnection, between science and decision-making. In this chapter, we start from the experience of the European Union with two foundations of risk regulation: evidence-based policy and the precautionary principle. The two are often contrasted for reasons of political advocacy, but – we will argue – they can coexist, at least within the sphere of the limits of freedom of scientific research. Our argument that evidence-based policy and the precautionary principle can and should be reconciled, however, is conditional, not absolute. We offer a proposition for public policy in which limits to the freedom of scientific research are removed in the name of the precautionary principle. Indeed, we argue that it is precautionary not to place limitations on the freedom of scientific research, because there are many possible adverse consequences for prosperity, innovation, welfare, health and society. At the same time, this lack of governmental intervention has to be balanced by a dialogic relationship between scientists and society. In the end our proposal is about a social contract: scientists obtain freedom but guarantee self-regulation and an active dialogue with society.

Governing risk in the European Union

In the experience of the EU, two foundations of risk regulation have emerged. Broadly speaking, we can call one foundation evidence-based policy and the other the precautionary principle. Evidence-based policy is, in principle, the main foundation of regulatory decision-making in the OECD countries and the EU. Evidence-based policy goes beyond risk regulation. Indeed, it is a cornerstone of the better regulation policy of the EU (European Commission 2015). It is a commitment to use evidence systematically in the life cycle of regulations. In fact, the Commission's regulatory policy is at least in principle anchored to evidence utilisation in the development of new EU legislation, risk analysis and ex-post legislative evaluation (European Commission 2015). This commitment to evidence on the part of the European institutions reflects a more general international trend. To illustrate, we find commitments to evidence-based policy in government guidelines for the development of laws and regulations, where the main policy instrument supporting public choice is regulatory impact assessment (Dunlop and Radaelli 2016).

Let us look at impact assessment because it provides a good example of how commitments to evidence-based policy operate at the level of decision-making. The thrust of impact assessment is to bring evidence to bear on regulatory choice: in taking a regulatory decision, for example whether to ban certain experiments with stem cells or the diffusion of genetically modified organisms, policymakers have to explain the reasons behind their choice, consult a wide range of stakeholders and – most relevant for our discussion – carry out an empirical analysis of the likely effects in terms of costs and benefits.

Thus, impact assessment has a core objective of identifying and possibly quantifying the likely effects of a proposed rule – imagine, for example, a regulatory proposal to set limitations to medical research or scientific experimentation. Typically, impact assessment revolves around the economic effects – costs and benefits – for different categories of stakeholders. But it can also look at intergenerational dynamics, the overall macroeconomic effects, the benefits and costs in terms of trade in open economies, CO_2 impact, demographic implications, income distribution and jobs. However, impact assessment embraces the evidence-based approach in a broader sense, comprising the obligation to state the reasoning behind regulatory intervention and to consult widely. In a nutshell, impact assessment and more generally evidence-based approaches to decision-making establish both rights and obligations: obligations for the regulators or lawmakers, and rights for those affected groups, professions and citizens who want to make their voices heard, and have the right to know about the empirical foundations of a regulatory proposal. The normative stance (i.e. what ought to happen, not necessarily what happens) of impact assessment is the following: in the absence of evidence and the possibility of discussing and criticising it, there is no social authorisation for regulation. Governments and institutions such as the European Commission cannot regulate unless they explain and illustrate empirically the reasons supporting regulation and allow for public comment. This normative stance is reflected in administrative procedure acts across the world, and therefore it governs administrative–regulatory interventions beyond the domain of impact assessment. As explained earlier, it is a manifestation of evidence-based policy as a foundation for public choice.

Let us now explore another aspect of this foundation. Regulations allow or prohibit certain types of behaviour. Some regulations restrict freedom of scientific research, often in the name of the environment, public health or ethical–religious principles. The evidence-based foundation for policy has a problem with these restrictions. Indeed, if we think from an evidence-based policy point of view, we should reason that regulators and policymakers have general duties – to protect the environment and public health, for example – but they need to be authorised via evidence-based tools each time they propose specific regulatory interventions.

Thus, there is something else (i.e. not evidence-based) that informs regulatory decisions. Here we come to another foundation of risk regulation: the precautionary principle. The conventional narrative, indeed, has pitched this principle against evidence-based policy. Before we critically consider this juxtaposition, we will clarify what the precautionary principle involves. In the context of the EU, the precautionary principle is enshrined (yet not defined) in Article 191 of the Treaty on the Functioning of the European Union of 2013 (TFEU). This article refers to the environment. It states that

> Union policy on the environment shall aim at a high level of protection taking into account the diversity of situations in the various regions of the Union. It shall be based on the precautionary principle and on the principles that preventive action should be taken, that environmental damage should as a priority be rectified at source and that the polluter should pay.

In 2000 the European Commission published guidelines on how to use the precautionary principle in a variety of policy domains. The precautionary principle 'applies where scientific evidence is insufficient, inconclusive or uncertain and preliminary scientific evaluation indicates that there are reasonable grounds for concern that the potentially dangerous effects on the environment, human, animal or plant health may be inconsistent with the high level of protection chosen by the EU' (European Commission 2000). In order to decide which kind of measure to take, there are some requirements. Indeed, any precautionary measure should be proportional, non-discriminatory, consistent with comparable measures already in place, be anchored to an examination of the benefits and costs of action and inaction, be subject to review and capable of assigning responsibility for producing the scientific evidence for a more comprehensive risk assessment (European Commission 2000: 3).

The precautionary principle should not be confused with prevention, where science 'can reliably assess and quantify risks' (COMEST 2005: 7), or pessimism, which is an inclination towards certain beliefs (Sandin 2004). We are dealing with risk management. The starting point is scientific uncertainty, which may be the consequence of either the need for further evidence (in certain cases total ignorance), or of the fact that we are dealing with trans-scientific issues, which are framed in the language of science but cannot be answered (at this moment in time or perhaps forever) by science (Weinberg 1972; Majone 2010). Although the definition of the principle is legal, it is political considerations that determine how and when constellations of actors invoke it (Tosun 2013).

Originally limited to environmental policies, the principle has been expanded by courts to public health and safety. There are traces of precautionary approaches in financial regulation too (Tosun 2013). In terms of regulatory philosophy, it is a foundational principle for the EU, like evidence-based policy. Given our word budget, we cannot rehearse the story of the principle, its possible interpretations (Gollier and Treich 2003) and its controversial applications to EU regulation (see Alemanno 2007: esp. chs II and III). We can, however, emphasise two points. First, as explained by Jale Tosun (2013), political considerations determine how the precautionary principle is invoked and triggered. A corollary of this political precautionary advocacy is the lack of transparency on how the interaction between precaution and evidence-based considerations is moulded in decision-making. We insist that the point is about the use of the principle, not something inherently political/opaque/anti-empirical in the principle itself. And in fact – our second point – the 2000 Communication of the European Commission (as well as regulation 178/2002 art. 6 on the European Food Safety Authority) anchors the principle to a set of requirements that are compatible with evidence-based policy – so much so that the Communication allows the EU regulators to trigger precaution only if the decision is based, among other things, on proportionality and benefit–cost considerations regarding intervention and inaction, and is subject to review in light of new scientific evidence. At least in legal and conceptual terms (if not in its usage), the precautionary principle is not incompatible with the other foundational principle of evidence-based decisions (Alemanno 2007). Unfortunately, we do not have sufficient case law regarding whether the conditions for triggering the precautionary principle have been met by the European Commission or EFSA (Alemanno 2007) – the European courts have been reluctant to provide a clear answer on this point.

A proposition for precaution to limit regulatory interventions, not to support them

As mentioned, it is political advocacy (not robust conceptual analysis) that has pitched the two foundations against one other: regulators either proceed on the basis of evidence or they invoke the precautionary principle. In Europe, we have seen this battle of principles being fought in many domains (European Risk Forum 2011; Garnett and Parsons 2016), from BSE (Monaghan et al. 2012: 181) to milk aids for cows, from regulation of medicines to chemicals (European Risk Forum 2016).

One of the best-known case concerns genetically modified food. Since the 1990s the European public has rejected GMOs, and authorities have implemented stringent regulations, 'sometimes citing vaguely' the precautionary principle without the support of strict scientific evidence. More complex social, ethical, cultural and economic factors were at stake (Wiener et al. 2011: 50). Recently, the use of bisphenol-A has been restricted, even though EFSA concluded that the evidence is too limited to draw any conclusions for human health. Another example is glyphosate. The European licence for the use of this substance has not been renewed yet (only extended for eighteen months), despite the findings of high-quality scientific assessments carried out by EFSA (European Risk Forum 2016: 34). Indeed, the European agency ruled that glyphosate is non-carcinogenic, drawing heavy criticism for its lack of data transparency and for being in contradiction of the IARC judgement that glyphosate is 'probably carcinogenic'.

So, are we predestined to follow one foundation of regulatory choice or the other? Not necessarily. First consider this: logically, there should be at least a minimum of empirical evidence to lead to the conclusion that 'we do not know enough' and opt for precaution. This is why in 2000 the Commission set evidence-based requirements for the use of the precautionary principle in decision-making. Second, consider the jurisprudence of the World Trade Organization: regulators cannot simply go for unqualified precaution, otherwise the use of the precautionary principle becomes equivalent to protectionism in disguise (Majone 2000). Third, at least in the EU, the principle is formally endorsed. It cannot be disposed of lightly. It has to be used with the other foundational principle of evidence-based policy, which is equally endorsed in all the strategic documents on regulation of the institutions of the EU. If the two are incompatible, we should conclude that the regulatory foundations of EU public policy contain two contradictory principles. Instead, as explained by Alemanno (2011), the two souls of EU risk regulation ought to improve their coexistence. Alemanno (2011) talks about humanising some features of risk analysis and asks for more transparency regarding how the two foundations come to play together in decision-making. This chimes with the debate on the other side of the Atlantic, where the Obama administration issued guidance on humanising cost–benefit analysis (Sunstein 2011).

But how exactly can the two foundations come together? We do not have a general answer to this question. But we suggest a solution in the special case where the core issue is freedom of scientific research. In this domain, we argue that the principle of precaution should survive, but (and this is the point) inside, not outside the empirical basis of decision-making. We therefore propose the following:

> Given that there is irreducible uncertainty in terms of technological risks and the economic and ethical issues caused by regulations that prohibit medical and scientific research, it is precautionary *not* to prohibit any scientific research unless there is empirical evidence showing that the costs and damage to people and environment outweigh the benefits of freedom of research.

This proposition takes an angle for the precautionary principle that has not been explored so far. The principle is typically used to regulate and ban, while we draw attention to domains where precaution suggests non-intervention. These domains are those where the key regulatory question is scientific research (as opposed to more applied stages such as innovation and the development of technology). To illustrate, we might think of regulations that prohibit research on assisted reproduction or on human embryos.

We hasten to add that our proposition is not unconditional. It is qualified by its own feasibility conditions. The first condition is that there has to be a certain maturity and institutionalisation of evidence-based practice in a given country or jurisdiction. Indeed, when we say 'unless there is empirical evidence', we mean that our argument is valid only if there is diffuse capacity to undertake empirical analysis. Evidence shows that the capacity for impact assessment, consultation and cost–benefit analysis differs widely across countries (Dunlop and Radaelli 2016).

The second condition is the social background that undergirds our formulation of the precautionary principle. Practically, we are thinking of a situation where science and society meet upstream, with several opportunities and instruments for public engagement at an early stage (Wilsdon and Willis 2004). The marriage of evidence-based policy and the precautionary principle does not materialise in a social vacuum. On the one hand, we need social trust. On the other, we need scientific responsibility on the part of the communities of science.

Social trust in science cannot be taken for granted. It has to be constantly produced and reproduced with appropriate forms of public engagement. Think of questions such as: what are the boundaries between basic and applied research? When does a scientific research project become a dangerous military application? How do we share the basic values of a certain development of medical research? In the past, governments have tried to 'educate the public' about science. This is the so-called deficit model (Ziman 1991; Sturgis and Allum 2004) in which the lack of trust in science is attributed to ignorance. Educate the masses, fill in their deficit of knowledge and there will be more trust in science. Today we know that scientific education, teaching statistics to journalists and other approaches have a very valuable role to play. But the reasons behind lack of trust are much deeper than ignorance.

At the same time, the solution cannot be limited to deontological codes. Our societies, and the EU especially, need a wider reconciliation between citizens and scientists. Let us consider all the dimensions of our equation. The government does not ban or limit scientific research unless there is compelling evidence of serious risk of harm to people and or the environment – this is the formulation of our principle. The scientists offer responsibility and engagement with society. Science becomes socially accountable, but not via governmental intrusion, regulations and, in the worst cases, faith-based obligations and prohibitions. All this amounts to a new role for the scientific community in society. This is possible if we identify a strong paradigm that generates self-regulation of scientists, accountability and dialogic attitudes.

Conclusion

This chapter offers a new definition of the precautionary principle, which goes hand in hand with the evidence-based policy in fostering freedom of research. We have argued that the precautionary principle should recommend non-intervention in scientific research, unless there is clear evidence showing that costs and damage outweigh benefits.

The challenge for the effective implementation of such a principle is first to create mature and institutionalised evidence-based practices. Yet this is not enough. Our version of the precautionary principle is effective only in a social environment characterised by social trust and scientific responsibility. How to build up this new 'social contract' between citizens, individual scientists, the scientific communities and the policymakers is the key issue for future research and public policy.

Note

1 We wish to thank Marco Cappato, Filomena Gallo and the Luca Coscioni Association for having invited Claudio Radaelli to present the initial draft of this chapter to the World Congress for Freedom of Scientific Research in Rome, 4–6 April 2014. We gratefully acknowledge the comments on the first draft provided by Alberto Alemanno, Lorenzo Allio, Claire Dunlop and Simona Giordano – the responsibility for mistakes and omissions remains ours.

References

Alemanno, A. (2007), *Trade in Food: Regulatory and Judicial Approaches in the EC and the WTO*, London: Cameron May.

Alemanno, A. (2011), 'Risk vs. hazard and the two souls of EU risk regulation: a reply to Ragnar Löfstedt', *European Journal of Risk Regulation* 2.2: 169–71.

COMEST (2005), *The Precautionary Principle*, Paris: UNESCO.

Dunlop, C. A., and Radaelli, C. M. (eds) (2016), *Handbook of Regulatory Impact Assessment*, Cheltenham: Edward Elgar.

European Commission (2000), *Communication on the Precautionary Principle* (COM 2000/001), Brussels.

European Commission (2015), *Communication from the Commission. Better Regulation for Better Results – An EU Agenda* (COM 2015/215), Brussels.

European Risk Forum (2011), *The Precautionary Principle. Application and Way Forward*, Brussels, www.riskforum.eu/uploads/2/5/7/1/25710097/erf_pp_way_forward_booklet_.pdf (last accessed 8 December 2016).

European Risk Forum (2016), *Scientific Evidence and the Management of Risk*, Brussels, www.riskforum.eu/uploads/2/5/7/1/25710097/erf_-_scientific_evidence___eu_risk_mgmt_16.pdf (last accessed 8 December 2016).

Garnett, K., and Parsons, D. J. (2016), 'Multi-case review of the application of the precautionary principle in European Union law and case law', *Risk Analysis*, 37.3: 502–16.

Gollier, C., and Treich, N. (2003), 'Decision-making under scientific uncertainty: the economics of the precautionary principle', *Journal of Risk and Uncertainty*, 27.1: 77–103.

Majone, G. (2000), 'The Credibility Crisis of Community Regulation', *JCMS: Journal of Common Market Studies*, 38: 273–302.

Majone, G. (2010), 'Foundations of risk regulation: science, decision-making, policy learning and institutional reform', *European Journal of Risk and Regulation*, 1.1: 5–19.

Monaghan, M., Pawson, R., and Wicker, K. (2012), 'The precautionary principle and evidence-based policy', *Evidence & Policy*, 8.2: 171–91.

Sandin, P. (2004), 'The precautionary principle and the concept of precaution', *Environmental Values*, 13: 461–75.

Sturgis, P., and Allum, N. (2004), 'Science in society: re-evaluating the deficit model of public attitudes', *Public Understanding of Science*, 13.1: 55–74.

Sunstein, C. R. (2011), 'Humanizing cost–benefit analysis', *European Journal of Risk and Regulation*, 2.1: 3–7.

Tosun, J. (2013), 'How the EU handles uncertain risks: understanding the role of the precautionary principle', *Journal of European Public Policy*, 20.10: 1517–28.

Weinberg, A. M. (1972), 'Science and trans-science', *Minerva*, 10.2: 209–22.

Wiener, J., Rogers, M., and Hammitt, J. et al. (2011), *The Reality of Precaution. Comparing Risk Regulation in the United States and Europe*, Washington, DC: RFF Press.

Wilsdon, J., and Willis, R. (2004), *See-through Science: Why Public Engagement Needs to Move Upstream*, London: Demos.

Ziman, J. (1991), 'Public understanding of science', *Technology & Human Values*, 16.1: 99–105.

15

Let freedom ring for science: an American perspective

Mary Woolley

Dr Martin Luther King's immortal phrase 'let freedom ring' is as thrilling today as it was when he first uttered it in 1963. Now, nearly half a century since the 1968 assassination of one of the most revered civil rights and moral leaders of our time, we celebrate Dr King's words as a touchstone and inspiration. With the famous march on Washington in 1963, Dr King attempted something extraordinary and the impact was enormous, driving social change and making an enduring difference in our society. Indeed, his successful strategy for achieving civil rights inspired millions of people worldwide, including patient advocates for research who also dreamed of changing the world. Since the time of Dr King, in the US and in many parts of the world, there have been many high-profile patient marches – and walks, runs and bicycle rides – organised to call attention and say 'enough'. People march in protest about ignoring or stigmatising HIV/AIDS; they march and rally to end breast cancer, Alzheimer's, diabetes, mental illness and many other diseases we don't have answers for yet, diseases for which research would give us answers, were it well supported.

The US political system responds to public protest, including protests expressed through marches, although it often takes a long time for a response to crystallise. I believe that if scientists, who are well respected by the public, joined patients in marching for research – both literally and through other means of stepping into the public arena – they would have great impact on political leaders. Unfortunately, scientists are rarely interested in doing so. It is not part of the culture of science to be active in engaging the non-science trained public. Not yet. But we are not going to have more science and more freedom of science, or more freedom of scientific enquiry, until the science community itself becomes more visible and engaged with the public and elected officials.

Based just outside Washington DC, Research!America is an alliance working to make the case for science. One example of our work is an advertising campaign intended to get the attention of the public, media and policymakers. The advertisement, which shows a warning label on a prescription medication container, says quite boldly: 'Washington politics just might kill you'. The advert generated a great deal of attention, allowing us to drive home the message that the inaction that characterises Congress then and now is not working to allow scientists to find the solutions to what ails us. Research!America works through social media as well. Social media is a very active, influential environment in which people learn and exchange information – good and bad information; accurate and inaccurate. Research!America is engaged in trying to make sure that people know where to get accurate information. In fact, we are working in every kind of media to make the case for research – truly the best promise of better health and well-being in the future for all of us. Unfortunately, that promise is too much at risk right now.

Science of all kinds is at risk in the US today. There have been cuts to federal government science funding for well over a decade. A breath of fresh air was provided in 2018 with increased funding for medical research, but the damage done by years of cuts has not been overcome. And there are problems beyond money. A number of elected officials threaten to shut down various kinds of science, including embryonic stem cell research and the social, behavioural and economic sciences. The tendency to either ignore or demonise science is made easier when the science community remains essentially invisible to political actors and the public. Largely because of this invisibility, US politicians rarely talk about science. Unlike in many other countries, and unlike in the US in the past, few elected officials in the US today have training in science. This can make them reluctant to talk about science. They are concerned that they might be asked a question they cannot answer. It is also a fact that they don't hear from many voters – their constituents – that research is important, and in general they are not being pressed for action. Complicating the situation is that, like many Americans, politicians often take scientific progress for granted. Because we have all seen a lot of scientific progress in our lifetimes, it is all too easy to overlook what it takes to assure that progress continues. But the problem is broader still.

In the US today, vocal groups are opposed to the government's role in science. In addition, there are many people, including elected members of Congress, who believe that we cannot afford to spend any more money on science. They say that the nation just does not have the money. Well, that is simply not true! There is a lot of money in the US; it is a matter of setting priorities as to how we choose to spend that money. For example, in 2017 pet owners in the US spent $70 billion on their pets (APPA 2017). Pets are wonderful, but the problem is that people don't see that with just a little more money devoted to science, not only they, but also their pets might have healthier lives. The $70 billion sum is about twice the budget of the entire National Institutes of Health (NIH). Of course, we don't advocate taking money spent on pets to put towards medical research. But using examples like this illustrates the point that money, per se, is not the problem we sometimes tell ourselves it is. We engage in false economies when we as a nation choose not to invest robustly in research. We are spending money on treating diseases such as Alzheimer's and many more, and at the same time we are starving research into those very diseases. One of the founders of Research!America, the philanthropist Mary Lasker said, 'If you think research is expensive, try disease.' She is right. What are we thinking? Why don't we spend more to prevent, eliminate or at least ameliorate disease and disability, rather than continuing to suffer physical, emotional and economic pain?

Research!America has one mission: to make research to improve health a high priority in the US. (We have four similar sister organisations in Sweden, Australia, Canada and New Zealand, and work with other like-minded organisations worldwide.) Founded in 1989, we are a not-for-profit alliance representing over 125 million Americans through their various organisations, universities, industry, patient groups and scientific societies. We are led by a magnificent volunteer board of well-respected individuals, many of whom have served in public office, as members of Congress and as leaders of federal agencies such as the NIH, the Centers for Disease Control and Prevention and the Food and Drug Administration. Our goals are to achieve more funding – we talk about money all the time! – and to ensure a positive, empowering policy environment that does not impede research in either the academic or the private sector. We make the case for public–private partnerships. We work to make sure that the public hears about research directly and via the media, and we also work to empower members of the science community to become effective spokespersons for research. Research!America's Chair Emeritus, a former member of Congress, the Honourable John Edward Porter, speaks often to members of the science community saying: 'You can change the image of things to come but you cannot do it sitting on your hands, you have to reach out to the Congress and build bridges.'

Scientists are often surprised to learn that most Americans cannot name a living scientist. In fact, 84 per cent say that they cannot name a scientist (Research!America, 2018). This shocking percentage is why I say that scientists are invisible in the US. Also troubling is that most Americans don't know where research is conducted: 66 per cent cannot give the name of a place where science is conducted (Research!America, 2018). President Abraham Lincoln, who established the US National Academy of Science in 1863, pointed out that 'public sentiment is everything. With public sentiment, nothing can fail; without it nothing can succeed'. We must take it to heart that earning and maintaining public sentiment – if you will, public trust – is essential. We can understand and track public sentiment via public opinion surveys. We know, for example, that the public supports basic research funded by the government; we know that the public supports STEM education – that is, science, technology, engineering and mathematics – and we know that the public believes that the government should make both science and STEM education a priority. However, some of the news we learn from commissioned surveys is not so good. People are split on their point of view as to whether the government should play a role in behavioural research; questions are raised as to whether the government should be involved in funding research to find answers to problems that involve personal choices, even when public safety and population health are at stake. If scientists were more visible and outspoken about the value of behavioural research, the public would be more inclined to listen. Public views do change, as they have about embryonic stem cell research. The public is now more likely to be positive than negative about this science, but it took a lot of work by many organisations and inspired individuals, including Bernard Siegel, to reverse what was once negative public opinion. It takes many people of conviction working together to speak out to influence public opinion, and with it policy and resource decisions. We should feel optimistic since history shows that advocacy works.

Importantly, we also know from surveys that the public say they are not well informed regarding what their elected representatives are doing and saying about medical and health research. In the 2016 U.S. elections, only 14 per cent of the public in fact said that they were very well informed of the positions of the candidates for president of the US regarding public policies and public funding for science and innovation, which is why we developed a programme that is used every election year to persuade those running for office to declare their views on research for health and science overall (Research!America 2016). We reach out to potential voters via social media and advertising to urge them to become informed (we don't take sides or back particular candidates). And we reach out to candidates to ask them to specify what their positions are regarding research. Our goal is to reach the point where anyone considering running for political office knows that she or he must articulate a position on medical research and science broadly, if they expect to be a successful candidate.

As our Chair Emeritus John Edward-Porter says, 'Wouldn't it be wonderful if all candidates had science advisers and advisory committees?' They would, if individual scientists step up and volunteer to be advisers to a candidate or an elective official. Physicist Dr John Holdren, who served as President Obama's scientific adviser, frequently remarks: 'Everybody in the science and technology community who cares about the future of the world should be tithing 10 per cent of his or her time to interacting with the public in the policy process . . . if all us just got out to the public more and talked to policymakers more, we would get more of this done.' What a difference that would make!

To again quote Dr Martin Luther King: 'Our lives begin to end the day we become silent about things that matter.' Science matters. It's time for scientists to overcome their reluctance to engage the non-science-trained public, addressing questions and talking about how they are serving the public's interest; describing how and why science matters to our society.

References

APPA (2017), www.americanpetproducts.org/press_industrytrends.asp (last accessed 25 May 2018).

Research!America (2018), National Public Opinion Survey, www.researchamerica.org/polls-and-publications (last accessed 25 May 2018).

Research!America (2016), National Public Opinion Survey, www.researchamerica.org/polls-and-publications (last accessed 25 May 2018).

Conclusion

Simona Giordano

This volume is the result of a long-standing international collaboration which takes the name of the World Congress for Freedom of Scientific Research (www.freedomofresearch.org). After the Third Meeting, held in Rome in April 2014, some of the presenters united in the preparation of this collection. Others interested in science and its regulation joined up along the way. The World Congress has overall brought together hundreds of people, including academics, policymakers, jurists, scientists and disability-right activists from all over Europe, the US, India, Iran and many other countries.

I will keep this conclusion brief, because what the authors have written already provides a great deal of food for thought. But I would like to offer a brief history of the World Congress, and to report some of the achievements obtained through this common international forum. The World Congress for Freedom of Scientific Research is a permanent forum of activities to promote freedom of scientific research worldwide; it was founded in Rome in 2004 by professor of economics Luca Coscioni. Professor Coscioni was diagnosed with amyotrophic lateral sclerosis. Within five years he was confined to a wheelchair, and founded, together with the main leaders of the Radical Party (Emma Bonino and Marco Pannella) a not-for-profit organisation that took his name: the Luca Coscioni Association. The World Congress for Freedom of Scientific Research was created to provide an arena for multidisciplinary and transnational discussion of scientific research, freedom and regulation, not attached to any specific political party. The First World Congress was held in Rome in 2006. At that time, Professor Coscioni was president of the Association, and was already severely ill. Sadly, he in fact died on the last day of the Congress. In a video message to the audience, Professor Coscioni said:

> The first meeting of the World Congress for Freedom of Scientific Research comes at a particularly difficult time in my life . . . Amyotrophic lateral sclerosis does not limit intellectual skills, it makes you fully aware of feelings

of despair and fear of lifetime. A time which is violently becoming narrower and which forces me to address the urgency of the price that millions of people around the world are paying and will have to pay to a culture of power, a culture of class . . . imbued with anti-scientific dogmas and prejudices, which exclude scientific knowledge and which exclude individual freedom to benefit from knowledge. Stakes are too high to let time pass, more time pass . . . To the violence of this cynical prohibition on scientific research and on the fundamental rights of citizens, I have responded with my body, which maybe many would have liked to see just as a hopeless prison, and today I respond with my thirst for air – because I am truly breathless – which is my thirst for truth, my thirst for freedom.

As Marco Cappato and I noted in the conclusion of our first volume on scientific freedom (Giordano et al. 2012), this message reminded us all that when we speak about scientific freedom we are not discussing an abstract idea: we are talking about real people, who have real lives and suffer real vulnerabilities and illnesses. I wish to add now that, as human life has extended so significantly in the last few decades, and as it is even clearer today than it was in 2006 that the process is not going to reverse or stop, it is imperative to remind ourselves that hope for treatment for many degenerative diseases (some of which are likely to come with longevity) bears upon scientific research, including stem cell research. The personal and social implications of scientific freedom and proper regulation is inestimable, as incalculable are the losses resulting from regulation that hinders or prohibits scientific progress unjustly, that bows to a culture not just of obscurantism, but even more worryingly of misinformation and conspiracy.

The first meeting of World Congresses for Freedom of Scientific Research was held in Rome, at the Campidoglio, the second at the European Parliament in Brussels, and the third again at the Campidoglio. The symbolic importance of this particular location needs to be stressed: the Campidoglio is in the very heart of Rome and very close to the Vatican. The square was designed and realised by Michelangelo Buonarroti, and the Campidoglio has been chosen as one of the symbols represented on Europe's coin, the euro. There is an interesting anecdote about this place. Before Christianity, the Campidoglio was the place where the pagans venerated the goddess Juno. Next to the goddess's temple were the 'sacred geese'. Around 390 BC Rome was besieged by the Gauls. It is said that on the night when the Gauls arrived, the sacred geese began to squawk so loudly that the consul woke up and alerted the city. According to the legend, it was Juno who woke up the geese – and after that Juno was also called Moneta, which in Latin means 'to warn, to caution' (*monere*). Incidentally, over one century later the mint was built near the temple, and the goddess Moneta was meant to protect the valuables. From this fusion of mythology and history the word *moneta* began to indicate currency – and hence *money* in English, *monnaie* in French, *moneda* in Spanish, *moeda* in Portuguese, *moneta* in Italian.

Perhaps now more than ever, particularly during the period of the separation of the United Kingdom from the European Union, with all this may imply for science (scientists' mobility, research funding, and so on) it is important to remember the significance of international collaboration – and therefore of forums of these kinds. Science is not just a human activity, but, arguably, a human right. As such science is described and defended in a number of international documents, declarations and covenants. The human right to science is enshrined in Article 27 of the 1948 Universal Declaration of Human Rights;[1] in Article 15 of the 1966 International Covenant on Economic, Social and Cultural Rights;[2] in Article 13 of the Charter of Fundamental Rights of the European Union;[3] in Article XIII of the American Declaration of Human Rights;[4] in Article 14 of the Additional Protocol to the American Convention on Human Rights in the Area of Economic, Social and Cultural Rights, the 'Protocol of San Salvador';[5] in Article 22 of the Charter of the African Union; in Part I of the Arab Charter of Human Rights; and in Article 32 of the Human Rights Declaration of the Association of Southeast Asia Nations.[6]

Specifically, Article 15 of the ICESCR sets forth:

1. the right of everyone:
   (a) To take part in cultural life;
   (b) To enjoy the benefits of scientific progress and its applications;
   (c) To benefit from the protection of the moral and material interests resulting from any scientific, literary or artistic production of which he is the author.
2. The steps to be taken by the States Parties to the present Covenant to achieve the full realization of this right shall include those necessary for the conservation, the development and the diffusion of science and culture.
3. The States Parties to the present Covenant undertake to respect the freedom indispensable for scientific research and creative activity.
4. The States Parties to the present Covenant recognize the benefits to be derived from the encouragement and development of international contacts and co-operation in the scientific and cultural fields. (www.ohchr.org/EN/ProfessionalInterest/Pages/CESCR.aspx)

This also means that citizens, all of us, are entitled to report specific violations of human rights, including those relating to science and the enjoyment of its benefits. The recognition of a human right is therefore also a call for mobilisation; it involves the acknowledgement of our responsibilities to check and strengthen the enjoyment of this right across the world.

It is obvious, and it will be obvious to the reader, that any area of scientific research needs to be regulated: clinical trials, the use of non-human animals in research, research on artificial intelligence and reproduction and so on are all regulated. Both financial constraints and regulations pose clear and inevitable limits to the freedom to perform scientific research. Knowledge transfer should be also regulated (as Rhodes persuasively argues in Chapter

6, this volume) because it is not always unproblematic. Freedom, thus, does not mean absence of regulation and lack of accountability. A serious ethical analysis of the various options, the advantages and disadvantages of various alternatives, the long-term consequences of scientific developments and innovations for future generations as well as for those living in other parts of the world, all needs to be part of dialogue on the regulation of research. This volume represents a contribution to this continued analysis. Although it is a small contribution, and certainly limited, I would like to stress the importance that enterprises such as these may bring to fruition by mentioning some of the achievements obtained through this forum.

The World Congress, together with fifty Nobel laureates who agreed to support the call, solicited the European Union to fund research involving embryonic stem cells. The 7th Framework Programme for Research and Technological Development was subsequently approved, which lifted the ban on the funding on embryonic stem cells, and thus now allows the funding of projects involving adult, induced pluripotent stem cells and embryonic stem cells, under special regulations and in accordance with national legislation. The World Congress also participated in action and consultation aimed at ensuring that this would continue in the 8th Framework Programme, now called Horizon 2020.

Another important achievement involved Costa Rica. In 2012 a hearing was held at the Inter-American Court of Human Rights concerning Costa Rican law which prohibited *in vitro* fertilisation. Two doctors from Costa Rica participated in one of the national meetings of the World Congress. The World Congress deposited a third-party judgment (amicus curiae) in defence of people's reproductive rights, highlighting the discriminatory nature of the law in question and its incompatibility with the fundamental human right to found a family.[7] Also as a result of the third-party intervention submitted by the World Congress, the courts condemned Costa Rica's legislation, which was consequently abrogated.

The World Congress for Freedom of Scientific Research has also presented or supported various petitions to the European Parliament, on euthanasia, HIV and reproductive and sexual health. The World Congress took an active role in the case of *Costa and Pavan* v. *Italy*. This case concerned a man and a woman in a relationship, both carriers of the gene for cystic fibrosis, who were denied access to pre-implantation genetic diagnosis under the Italian Law 40 on assisted fertilisation.[8] The World Congress submitted an amicus curiae with various patients' coalitions and sixty MPs. On 28 August 2012 the European Court of Human Rights condemned Italy for violation of Article 8 (right to respect for private and family life) of the European Convention on Human Rights.

The World Congress has, between 2016 and 2017, repeatedly denounced to the United Nations the repressive policies adopted in the Philippines by Rodrigo Duterte, under the guise of the 'war on drugs'. As may be known, the anti-drug hard line taken by Duterte hits the poorest sections of society,

and has resulted in the killing of over 8,000 people. R
World Congress participated and presented evidence a
the UN Commission on Narcotic Drugs held in Vie
also in support of the vice-president of the Philippines, I
Robredo, who has repeatedly intervened to denounce t
by Duterte in the name of the war on drugs.

These actions and achievements are important, as they illustrate the
of coordinated work between democratic states for the active consolidation
of human, civil and political rights worldwide. Moreover, they also illustrate
that the activation of international jurisdictions can guarantee supranational
protection of democratic rights. Forums such as the World Congress ensure
that debates do not remain confined to the 'ivory tower' of 'armchair academics', or within specific disciplines, or within specific ideological alliances,
but can result in practical and political initiatives that remain sensitive to
cultural identities.

The main message that this volume wants to convey is that science,
research, development and the protection of the human, civil and political
rights of many of us depend on the cooperation of many: scientists and
policymakers of course, but also academics and lay citizens. Scientific
development also depends on continued dialogue with all political parties;
it involves popular action at national and international levels, consultations,
knowledge transfer and, it is important to stress, non-violent action. Strong
ideological opposition in delicate areas such as embryo research, narcotic
drug use, women's health (fertility treatments of various kinds) has fomented,
as is sadly well known, episodes of serious violence. This book stresses the
importance of dialogue and non-violent mobilisation.

This volume is perhaps only a small contribution to the international
debate on freedom of scientific research, and there are many areas of science
that we have not considered – and many ideas about regulation that we
have not explored. But we hope that it will provide a method of cultural
exchange, give some interesting perspectives and stimulate further debates
on issues relating to science, freedom of research and individual rights and
responsibilities.

Notes

1 www.un.org/en/universal-declaration-human-rights (last accessed 26 October
  2017).
2 www.ohchr.org/EN/ProfessionalInterest/Pages/CESCR.aspx (last accessed 26
  October 2017).
3 www.europarl.europa.eu/charter/pdf/text_en.pdf (last accessed 26 October 2017).
4 www.narf.org/wordpress/wp-content/uploads/2015/09/2016oas-declaration-
  indigenous-people.pdf (last accessed 26 October 2017).
5 www.oas.org/juridico/english/treaties/a-52.html (last accessed 26 October 2017).
6 www.asean.org/storage/images/ASEAN_RTK_2014/6_AHRD_Booklet.pdf (last
  accessed 26 October 2017).

ww.associazionelucacoscioni.it/wp-content/uploads/2012/11/atto-Definitivo-intervento-paginado-traducido-_4_.pdf (last accessed 26 October 2017).
8 See official documents at https://strasbourgobservers.com/category/cases/costa-and-pavan-v-italy (last accessed 26 October 2017).

Reference

Giordano, S., Coggon, J., and Cappato, M. (eds) (2012), *Scientific Freedom*, London: Bloomsbury.

Index